Sponsored by the
European Association of Neurosurgical Societies

Advances and Technical Standards in Neurosurgery

Edited by

L. Symon, London (Editor-in-Chief)
J. Brihaye, Brussels
F. Cohadon, Bordeaux
B. Guidetti, Rome
F. Loew, Homburg/Saar
J. D. Miller, Edinburgh
H. Nornes, Oslo
E. Pásztor, Budapest
B. Pertuiset, Paris
M. G. Yaşargil, Zurich

Volume 15

Springer-Verlag

Wien New York 1987

With 36 Figures

Product Liability: The publisher can give no guarantee for information about drug dosage and application thereof contained in this book. In every individual case the respective user must check its accuracy by consulting other pharmaceutical literature.

Library of Congress Catalog Card Number 74-10499

ISBN 978-3-7091-7461-6 ISBN 978-3-7091-6984-1(eBook)
DOI 10.1007/978-3-7091-6984-1

ISSN 0095-4829

Preface

As an addition to the European postgraduate training system for young neurosurgeons we began to publish in 1974 this series devoted to Advances and Technical Standards in Neurosurgery which was later sponsored by the European Association of Neurosurgical Societies.

The fact that the English language is well on the way to becoming the international medium at European scientific conferences is a great asset in terms of mutual understanding. Therefore we have decided to publish all contributions in English, regardless of the native language of the authors.

All contributions are submitted to the entire editorial board before publication of any volume.

Our series is not intended to compete with the publications of original scientific papers in other neurosurgical journals. Our intention is, rather, to present fields of neurosurgery and related areas in which important recent advances have been made. The contributions are written by specialists in the given fields and constitute the first part of each volume.

In the second part of each volume, we publish detailed descriptions of standard operative procedures, furnished by experienced clinicians; in these articles the authors describe the techniques they employ and explain the advantages, difficulties and risks involved in the various procedures. This part is intended primarily to assist young neurosurgeons in their postgraduate training. However, we are convinced that it will also be useful to experienced, fully trained neurosurgeons.

The descriptions of standard operative procedures are a novel feature of our series. We intend that this section should make available the findings of European neurosurgeons, published perhaps in less familiar languages, to neurosurgeons beyond the boundaries of the authors countries and of Europe. We will however from time to time bring to the notice of our European colleagues, operative procedures from colleagues in the United States and Japan, who have developed techniques which may now be regarded as standard. Our aim throughout is to promote contacts among neurosurgeons in Europe and throughout the world neurosurgical community in general.

We hope therefore that surgeons not only in Europe, but throughout the world will profit by this series of Advances and Technical Standards in Neurosurgery.

The Editors

Contents

Listed in Index Medicus

A. Advances

B. Technical Standards

List of Contributors

Drake, C. G., M.D., F.R.C.S.C., Division of Neurosurgery, University Hospital, 339 Windermere Road, London, Ontario N6A4 5A5, Canada.

Gur, D., Sc.D., Department of Radiological Imaging, University of Pittsburgh, Pennsylvania, U.S.A.

Gybels, J., M.D., Kliniek voor Neurologie en Neurochirurgie, Universitair Ziekenhuis "Gasthuisberg", B-3000 Leuven, Belgium.

Jewkes, Doreen, M.D., Department of Neurological Surgery, Institute of Neurology, The National Hospital, Queen Square, London WC1N 3BG, U.K.

Latchaw, R., M.D., Department of Radiology, Presbyterian-University Hospital, Pittsburgh, Pennsylvania, U.S.A.

Nemoto, S., M.D., Division of Neurosurgery, University Hospital, 339 Windermere Road, London, Ontario N6A4 5A5, Canada.

Peerless, S. J., M.D., Division of Neurosurgery, University Hospital, 339 Windermere Road, London, Ontario N6A4 5A5, Canada.

Van Roost, Dr. D., Neurochirurgische Universitätsklinik, D-5300 Bonn-Venusberg, Federal Republic of Germany.

Thomas, Dr. D. G. T., Department of Neurological Surgery, Institute of Neurology, The National Hospital, Queen Square, London WC1N 3BG, U.K.

Wolfson, Jr., S. K., M.D., Department of Neurological Surgery, Montefiore Hospital, Pittsburgh, Pennsylvania, U.S.A.

Woolf, C. J., M.D., Cerebral Functions Research Group, Department of Anatomy, University College London, Gower Street, London WC1E 6BT, U.K.

Yonas, H., M.D., Department of Neurological Surgery, University of Pittsburgh, Pennsylvania, U.S.A.

A. Advances

Stable Xenon CT/CBF Imaging: Laboratory and Clinical Experience

HOWARD YONAS

Department of Neurological Surgery, University of Pittsburgh, Pennsylvania (U.S.A.)

DAVID GUR

Department of Radiological Imaging, University of Pittsburgh, Pennsylvania (U.S.A.)

RICHARD LATCHAW

Department of Radiology, Presbyterian University Hospital, Pittsburgh, Pennsylvania (U.S.A.)

SIDNEY K. WOLFSON, JR.

Department of Neurological Surgery, Montefiore Hospital, Pittsburgh, Pennsylvania (U.S.A.)

With 18 Figures

Contents

Introduction

Since Winkler in 1977 determined that the radiodensity of xenon could be used for image enhancement during transmission computerized tomography[44], a number of groups have explored the possibility of using stable xenon as a tracer of cerebral blood flow (CBF). In 1978, Kelcz and coworkers[25] defined the CT enhancement characteristics of stable xenon at the same time that Drayer, *et al.*[6] reported early experiments with the direct measurement of CBF. Since that time, work has continued in the search to determine the best route by which to characterize the time course of arterial and tisssue enhancement, as well as calculate CBF using this new technique.

Published work in the last several years has addressed the applications of xenon/CT CBF imaging to a spectrum of clinical disorders[3, 8, 9, 21–23, 30–33, 40, 42, 43, 49, 53, 54, 56]. In 1984, our group at the University of Pittsburgh, working with researchers from the General Electric Company, devised a means of integrating the xenon/CT CBF methodology within the GE/9800 CT system. Efforts by other companies suggest that the method will be incorporated into many other CT systems in the near future.

Methodological Considerations

The arterial concentration of xenon was initially analyzed by Drayer *et al.*, by scanning samples of arterial blood withdrawn during xenon inhalation. While some groups have continued to use the scanning of externalized arterial blood, others have adopted a "shuttle" technique that involves the scanning of the cervical carotid artery between scans of intracranial structures. Another approach involves the use of end-tidal xenon concentration to provide, indirectly, a measure of arterial build-up of xenon concentration. While a mass spectrometer has been used for this purpose[4], the thermoconductivity analyzer has also proven to be a sensitive and stable device[15, 16] with potentially similar accuracy as a mass spectrometer[17].

To characterize the movement of xenon within the brain it was apparent to us that, because the stable xenon and CT imaging procedure requires a limited number of CT scans obtained at relatively long intervals (20+ seconds), a "curve stripping" approach such as that used with xenon-133 CBF determinations could not be used with stable xenon and CT imaging. As a result, most investigators were forced to assume that each computed voxel (unit volume), which could be as small as $0.5 \times 0.5 \times 5\,mm^3$, consisted of a single flow compartment[7, 13, 29]. This assumption also meant that the "build-up curve" could be used for CBF analysis. The high resolution and stability of new scanners has made these assumptions increasingly more valid[16, 26].

While other CBF methodologies have used normative partition coefficient values, we have found that the partition coefficient (lambda) can be directly measured by using stable xenon in combination with imaging until all brain structures are saturated with xenon. Gray matter saturation normally occurs within three minutes; white matter saturation can require 20 to 30 minutes[5, 7, 29]. Currently, mathematical means of extrapolating a lambda value from a relatively brief period of inhalation measuring 4.5–5 minutes are being employed[16]. This approach has proven to provide reasonable flow information within the scope of a more clinically acceptable examination.

Although some groups have calculated CBF using a single enhancement point and others have assumed a "normative lambda"[6, 11, 29], a two-variable approach using the best fit for the data has significantly reduced errors associated with the xenon/CT CBF calculation[16, 19, 26, 41]. The two-variable approach and modern computer applications have enabled the calculation of CBF for every CT voxel[16, 39], resulting in the creation of maps of flow with relatively high resolution.

A major clinical limitation of xenon/CT CBF measurement is the marginal signal-to-noise ratio that results when lower, more clinically acceptable concentrations of xenon are used[15, 50, 51]. While 100% xenon provides about 30 Hounsfield units of tissue enhancement, the inhalation of this concentration for even a single breath can be dangerous[46]. Concentrations of 50–60% consistently cause sedation and have been associated with bronchospasm[36]. While initial observations suggested that concentrations ranging from 24–50% could be tolerated, our experience with more than 1,800 patients has led us to conclude that concentrations of xenon at or above 35% are associated with an unacceptably high incidence of patient intolerance. We have, however, observed that at an inhaled concentration of 28–33% xenon, most individuals experience few ill effects[28]. In recent studies we have observed that the addition of 0.8% CO_2 to a mixture of 32.2% xenon and 67% oxygen further stabilizes respiratory parameters[1]. An acceptable signal-to-noise ratio has been made possible with these lower xenon concentrations because of the comparatively high signal-to-noise ratio and image stability provided by the newer CT scanners including the GE/9800.

Clinical Methodology

The xenon CBF imaging methodology that was developed at the University of Pittsburgh and subsequently incorporated into the GE/9800 system will be reviewed here. The system is now commercially available and is being used in over 70 centers worldwide. Collectively, these centers have conducted more than 3,000 studies in which our imaging procedure has proven to be a useful and clinically acceptable approach to the noninvasive acquisition of meaningful local cerebral blood flow (LCBF) data[28].

Although basic CT technology is not inexpensive, the additional expense required to make CBF imaging possible is comparatively small and the equipment need not be dedicated to blood flow measurements. In our institution this has resulted in optimum use of scanning equipment, with flow studies now being routinely performed within the time frame required for a normal baseline and iodine enhanced CT study. A typical study proceeds as follows;

Xe/CT Study Procedure

Before being positioned on the CT table, the patient is informed that transient sensory disturbances may occur. He/she is told that it will be necessary to remain completely still throughout the procedure, which involves two baseline scans and by $4^1/_2$ minutes of xenon inhalation for a total period of about $6^1/_2$ minutes. Blood pressure is recorded at this time and immediately after the examination.

The patient is then positioned within the CT scanner and fitted within a standard CT scanning headholder combined with an attached vacuum device (Vac-Pac, Olympic Medical Supply Co., Seattle, WA) to aid fixation of the head. A limited number of initial scans are taken to select the levels for CBF analysis. Because CT artifacts significantly degrade the flow information, CT scan angles are selected that will minimize this artifact. The sphenoid and petrous bone as well as the midline frontal and occipital protuberances are intentionally avoided. Exposure of the orbits is also avoided, to reduce the radiation exposure to the radiosensitive lens of the eye.

The patient is reminded once more to remain completely still during the study. He/she is then fitted with either a face mask or a mouthpiece (accompanied by firm nasal compressions) and the xenon gas mixture (32.2% xenon, 67% oxygen, 0.8% CO_2) is delivered from a 60-liter plastic bag through a nonrebreathing system. End-tidal pCO_2 is measured with a capnograph, and the xenon concentration is monitored by a thermo-conductivity detector. Both of these measurements are continuously displayed. Two baseline scans are then obtained at each level of study. The patient's tolerance for xenon gas is monitored throughout the examination by predetermined toe movements that communicate with the clinican.

We study multiple brain levels during one inhalation period by using programmed incremental table movements between four-second CT scans obtained at 60-second intervals at each level[14]. Up to three levels can be selected for study during one inhalation sequence, with the number of levels limited at this time by the heat-loading constraints of the current CT X-ray tube.

Xenon end-tidal values obtained from the thermoconductivity detector (Fig. 1) and the recorded percent xenon are converted into an estimated time-dependent CT enhancement of the arterial blood using the following relationship:

Ca(u) = C Xe ET (1 + 1.1 Hct.),
Ca(u) = instantaneous arterial Xe concentration,
C = kVp dependent conversion constant,
Xe ET = end-tidal Xe concentrations,
Hct = hematocrit.

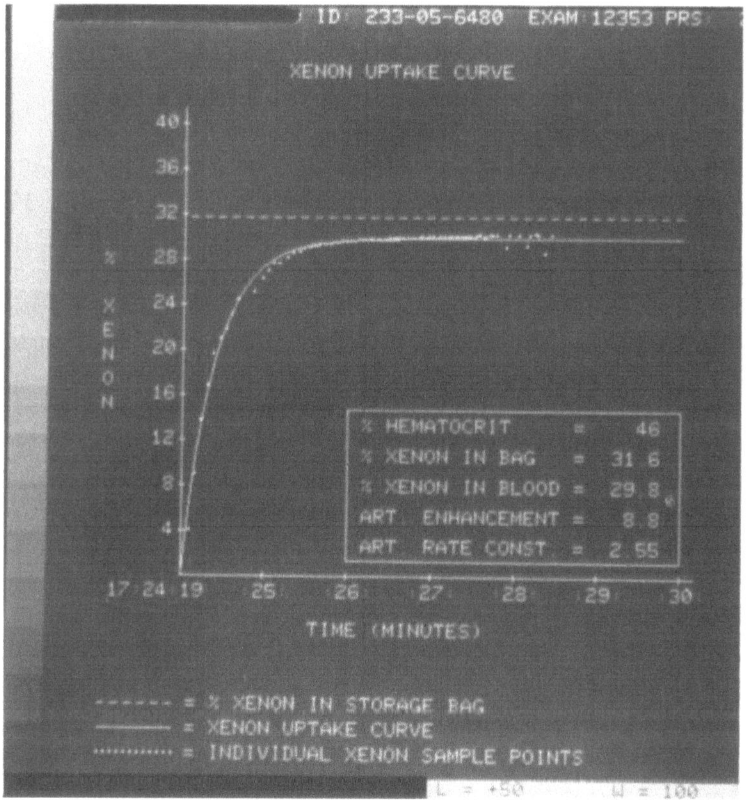

Fig. 1. The xenon uptake air curve is displayed on a scale of percent xenon versus time in minutes. Each end-tidal xenon value is displayed as a dot and the initial concentration of xenon in the reservoir is recorded with a "dashed" line. The exponential of the fitted rate constant is displayed as a solid line. In the study data box on the lower right, the first entry is the patient's premeasured hematocrit recorded as a percentage. The % xenon in the bag is the next entry, and the maximum % xenon in blood as measured from the fitted rate constant is the third. The fourth number is the arterial enhancement, which is calculated from the hematocrit and the maximum % xenon in blood (see text). Lastly, the arterial rate constant is the exponential fitted rate constant in units of $(\text{min})^{-1}$ at which blood approaches xenon saturation

The sequence of CT images obtained at each brain level before and during xenon inhalation is used to characterize the local build-up of xenon in tissue for the calculation of LCBF. The two baseline scans obtained before xenon inhalation are averaged to reduce noice level, and this averaged baseline image is subtracted from each image obtained during xenon inhalation. Each voxel measuring $1 \times 1 \times 5$ or $1 \times 1 \times 10\,mm^3$, is subsequently defined by a series of delta CT values as a function of time. These data are combined with the measured arterial concentrations over time to solve the Kety/Schmidt equation.

$$\Delta CT(t) = f \int_0^t \Delta Ca(u)\, e^{-k(t-u)}\, du \qquad \text{where } f = k \cdot \lambda$$

A weighted least-squares fit routine is used to derive estimates of the two parameters, flow (f) and lambda (λ) using an iterative approach. Pre- and postanalysis smoothing routines are used to reduce pixel-to-pixel noise. A centrally weighted, three-pixel, bell-shaped smoothing routine is used before the computation, and a centrally weighted 9×9 or 7×7 bell-shaped smoothing routine is used after the computation. When a point data set cannot be fit to the xenon build-up curve within the limits of allowable error (ssq), the system automatically averages values from the adjacent eight voxels and substitutes that value.

Calculation of flow values for each of the 10,000 to 30,000 voxels per CT level requires 14 minutes, resulting in a flow map (Fig. 2). Quantitative data are presented on the CT console adjacent to the baseline CT image. In addition, a "confidence image" is presented which represents a normalized sum of the square of the differences between the actual enhancement values and the fitted exponential of enhancement values. The degree of whiteness on the confidence image correlates with a poorer fit and greater error inherent in that computed CBF value.

The flow images can be analyzed in a number of ways. With the large number of flow calculations acquired, the mean and standard deviation of flow value for any desired region can be displayed. For example, in Fig. 3 a, a continuous quantitative gray scale is used to display the flow data. In this map, the average flow value of the 16 two-centimeters-in-diameter regions of interest (ROIs) was $49\,cm^3/100\,gm^3$/minute with values for each ROI being dependent both on arbitrary placement and on the mix of tissue within each. The selective placement of 5 mm ROIs within anatomical gray matter areas of the CT image provided values from $41–129\,cm^3/100\,gm^3$/minute (Fig. 3 b), with a mean of $74.8\,cm^3/100\,gm^3$/minute; the associated white matter mean for similarly generated regions of interest was $19.2\,cm^3/100\,gm^3$/minute. The gray/white ratio in this normal study is 3.9.

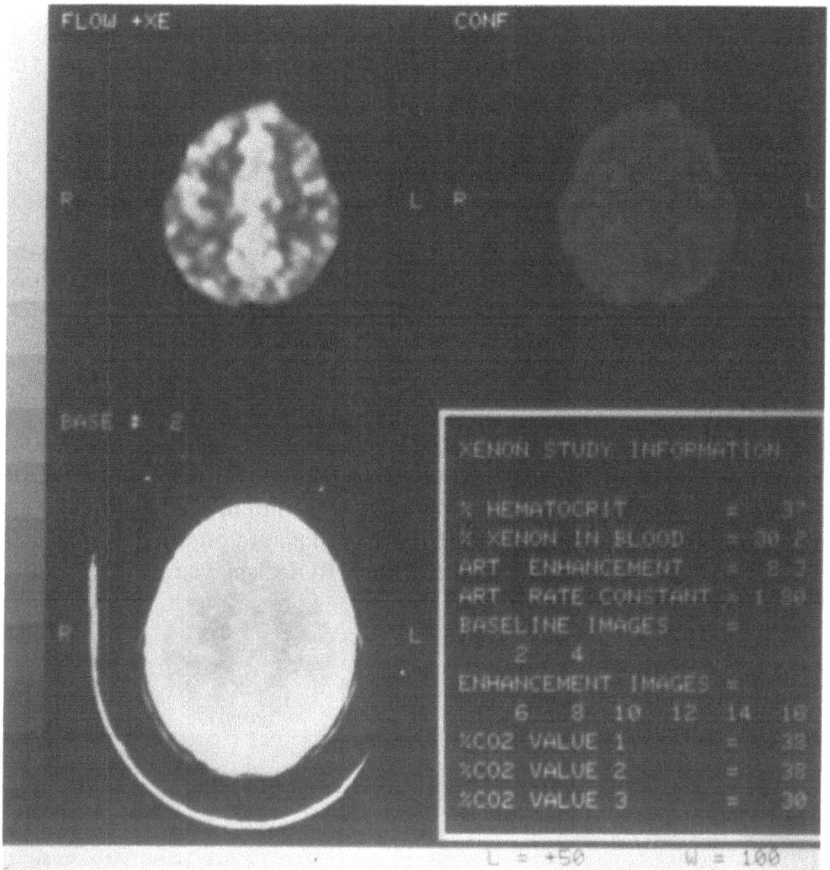

Fig. 2. The smoothed flow image is displayed on the upper left on a gray scale of 0–100 cm^3/100 g/minute when the window level (*w*) below the image on the right is set at 100 and the level (*L*) is set at 50. The baseline scan for the flow map is displayed below the flow image with the confidence image on its right. In the "xenon study information" box, the baseline and enhanced images used in the calculation are displayed based on the data already described in Fig. 1. Lastly, pCO_2 levels are recorded at the beginning, middle, and end of the period of xenon inhalation

A number of groups working with the stable xenon/CT method have also shown values consistent with established flow norms[7, 29, 39].

Laboratory Validation

Over the past nine years, we have examined in animal studies the possible physiological significance of the blood flow information provided by this methodology. Flow resolution as defined by full width half-maximum

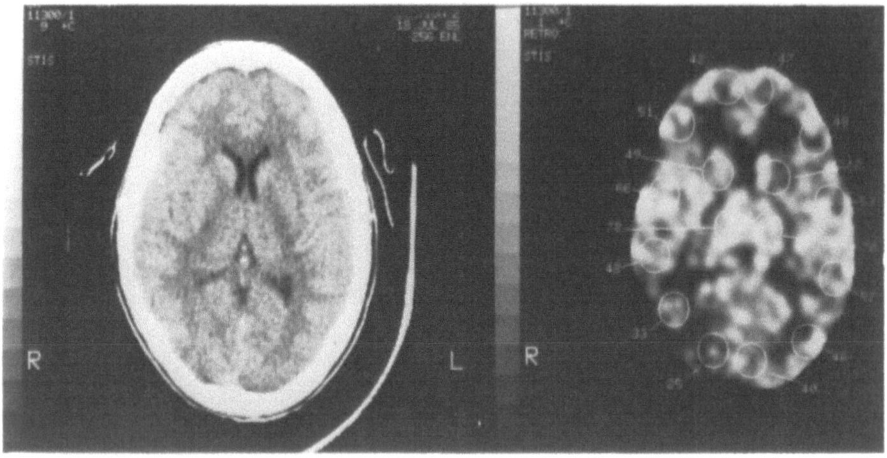

Fig. 3 a. This is a relatively normal study with the baseline CT displayed on the left and its accompanying flow map on a gray scale of 0–100 cm^3/100 g/minute on the right. Two-centimeters-in-diameter ROIs have been placed on the flow image and the average flow value of each ROI displayed

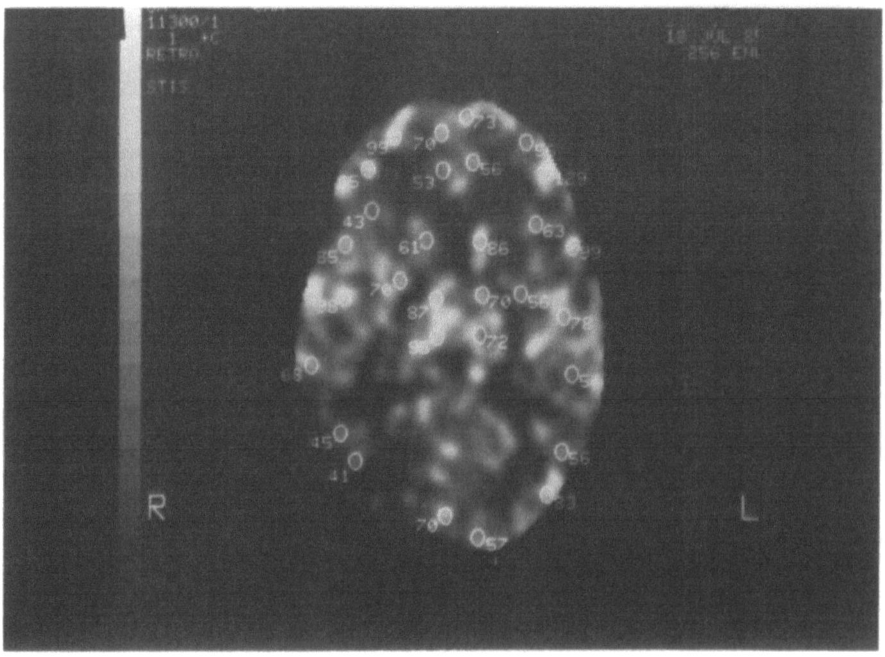

Fig. 3 b. The scale is 0–150 cm^3/100 g/minute, and five-millimeters-in-diameter ROIs, which were selected from within gray matter regions on the CT image, are displayed

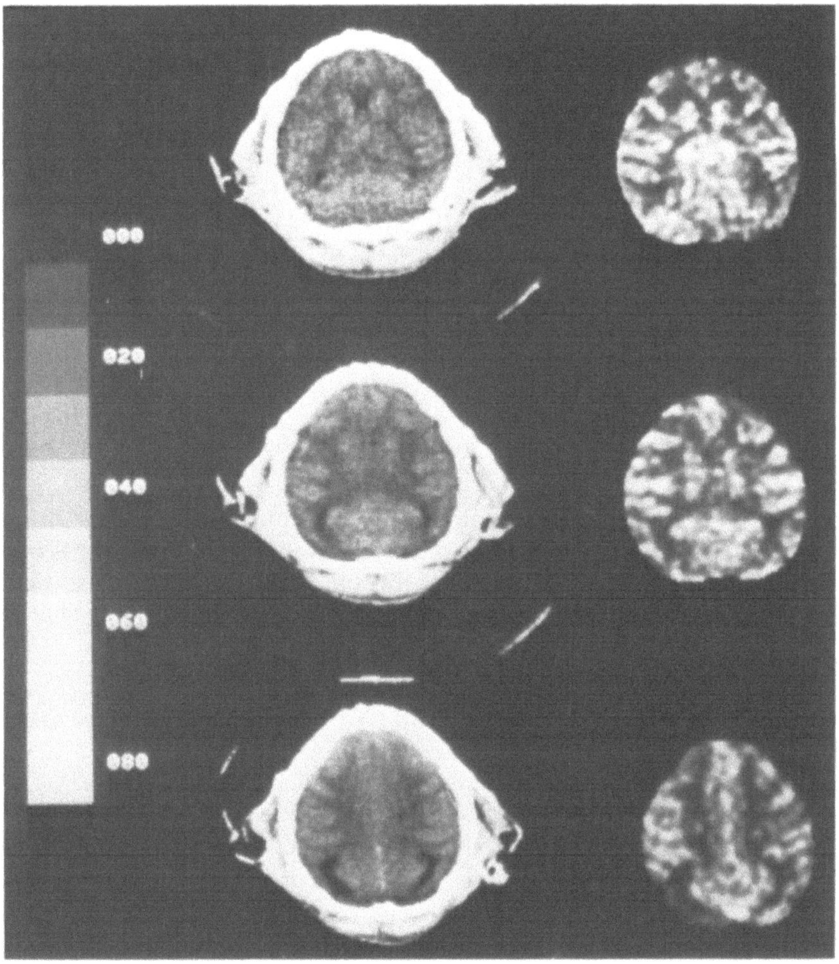

Fig. 4. This normal three-level blood flow study was obtained in a baboon under neuroleptic anesthesia with a pCO_2 of 44 mm Hg. Note the high degree of correlation between the adjacent anatomy (as seen on the CT image) and appropriate regions of high and low flow consistent with gray and white matter

(FWHM) is approximately 6 mm with the current commercially available system. It can be improved somewhat ($\sim$ 4.5 mm) when small brains are studied. The ability of the current methodology to resolve variations of flow between gray and white matter has been shown in our animal studies, in which ribbons of gray matter measuring only 2 mm and the internal capsule measuring $\sim$ 2–3 mm are readily evident[13,16] (Fig. 4).

To further validate these observations, we conducted a series of microsphere experiments in which radiolabeled microsphere flow measure-

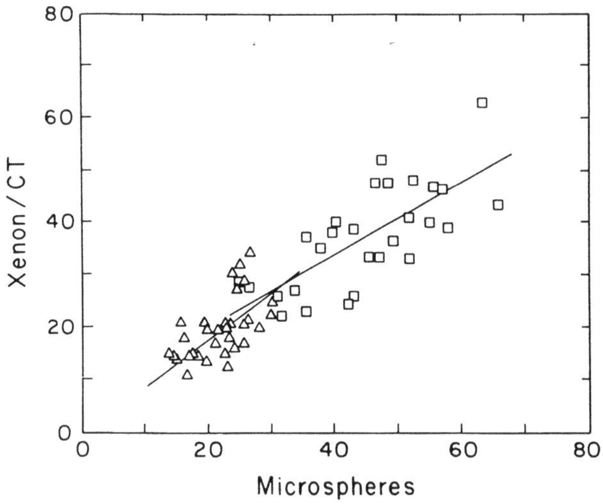

Fig. 5. This is a plot of concurrent flow values found with xenon/CT and radiolabeled microsphere CBF measurements. (Permission, Gur, Ref. 15.) A basic difference in methodologies is suggested by the relatively higher flow values recorded with microspheres with high flow. Units are 100/100 gms/min

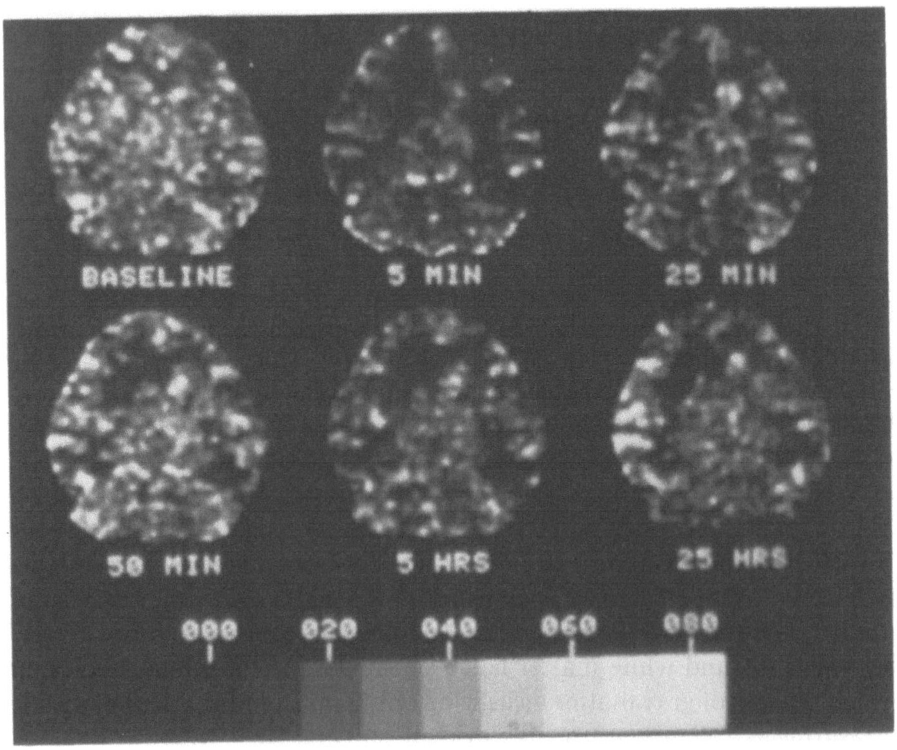

Fig. 6 a. Transorbital coagulation of the lateral lenticulostriate arteries produced a permanent absence of CBF within the distribution of those vessels, in addition to an early global reduction of flow. Note the occurrence of more subtle regional alterations of flow over time within the ipsilateral and contralateral hemispheres. The gray scale is 0–80 cm^3/100 g/minute

ments were made before, during, and after stable xenon inhalation. A high degree of correlation was identified among these studies for both high and low flow (Fig. 5)[19]. A similar study with similar results was recently reported by Panos[35].

The reproducibility of the information obtained from the Xe/CT method has been shown in repetitive studies in animals to be approximately 12% for ROIs 1.4 cm in diameter or larger. Due to system noise, smaller ROIs have a higher variation between measurements, with the information provided by a single voxel ($1 \times 1 \times 10\,mm^3$) having a flow-dependent standard deviation (higher with low flow) which may be as large as 100%[12].

We and others have performed studies involving the manipulation of CO_2 and blood pressure to evaluate whether the presence of xenon altered these physiological responses[16,27,48]. In both animal and human studies, CO_2 manipulation has been shown to cause an alteration of CBF which ranges from 3–5% for each millimeter of Hg CO_2 variation in normal subjects to as

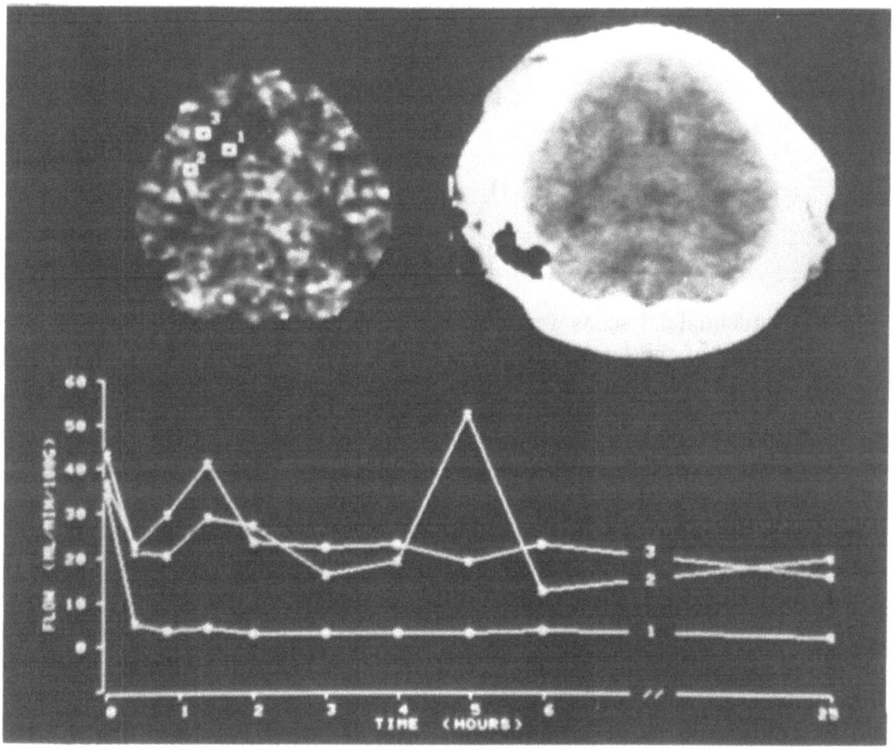

Fig. 6 b. Displays flow values obtained over time from two-centimeters-square ROIs placed within the right caudate (*1*) and the adjacent rim of the lentiform nucleus (*2, 3*). Note the stability of flow values within the caudate over time and the variability of flow alterations in the immediately adjacent tissues

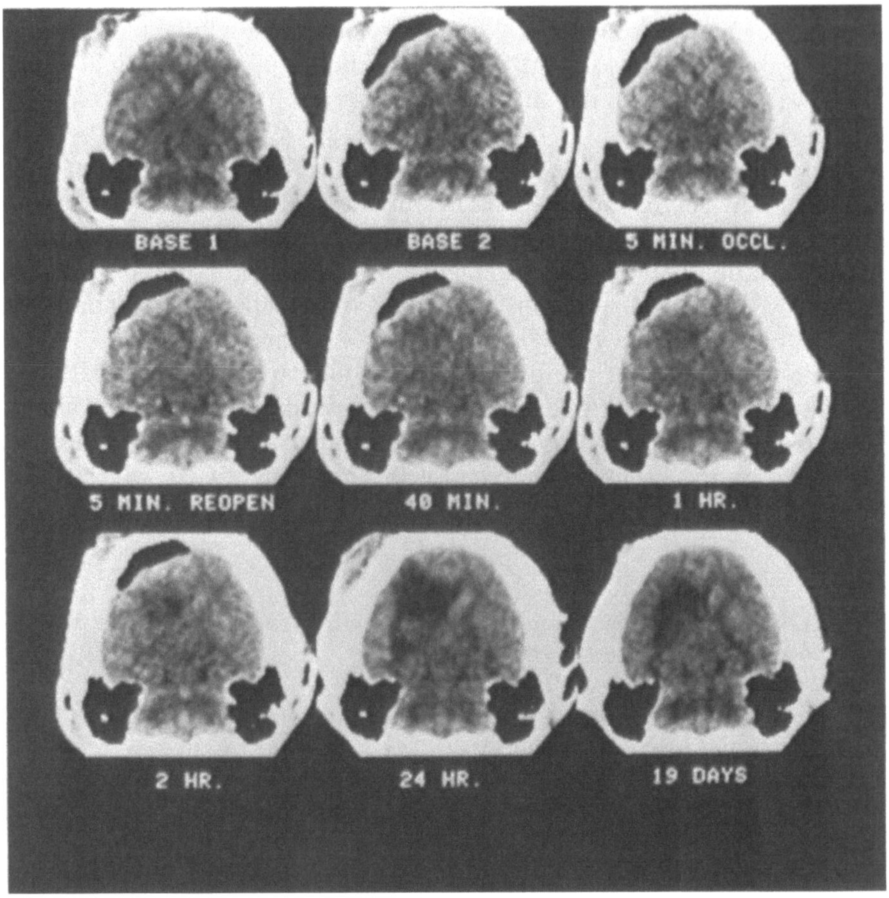

Fig. 7 a. Sequential CT scans were obtained prior to, during, and following a one-hour occlusion of the lenticulostriate arteries. Note the appearance of CT low density within two hours after vessel reopening and the progression to encephalomalacia by 19 days

low as 1% in subjects who have undergone severe closed-head injury and have massive swelling. Experimental studies involving hypotension due to controlled blood loss have consistently demonstrated the stability of blood flow and presumably the integrity of autoregulation within a normal range of blood pressures until mean arterial pressures below 30–40 mm of Hg are reached[48].

In studies of focal ischemia in animals, we have used the Xe/CT technique to examine occlusion of perforating vessels (lateral lenticulostriate) occlusion[50]. Selective occlusion of only the lenticulostriate arterials has been seen to produce a region with near zero flow in the ipsilateral caudate and putamen, as well as in the intervening internal capsule

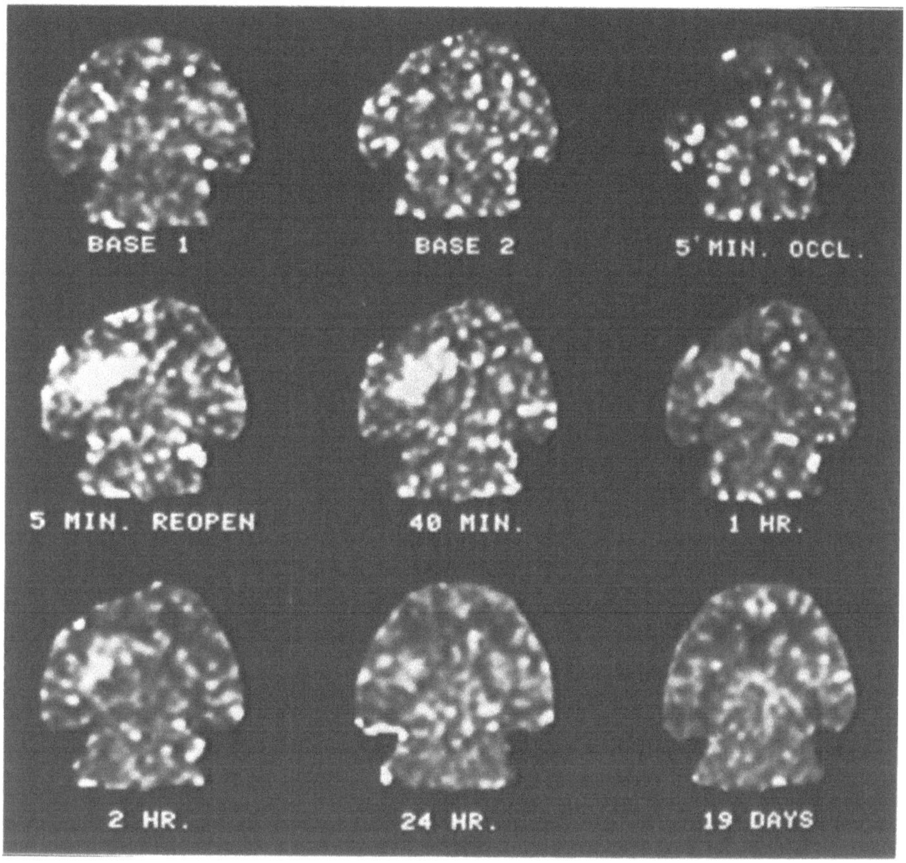

Fig. 7 b. Lenticulostriate occlusion again produced a dramatic reduction of local CBF (< 1 cm^3/100 g/minute) within the distribution of these vessels. Reopening the lenticulostriate vessels resulted in an early hyperemia with focal flows of 80–100 cm^3/100 g/minute. Flow values returned to low flow within 24 hours of reperfusion

(Fig. 6 a)[55]. In every experimental study we have performed in which less than 1 cm^3/100 g/minute has been recorded for one or more hours, neuronal death has been apparent on pathological examination at six hours postocclusion, with the region proceeding in other animals to encephalomalacia on pathology examination at three to four weeks. The reproducibility and stability of the Xe/CT is evident from the stability of flow values displayed in Fig. 6 b, in which blood flow values obtained from identical regions over a 25-hour period are displayed from basal ganglion ROIs. In other studies, reperfusion of the vascular bed after one hour of occlusion, accomplished by removal of a clip on the lenticulostriate arteries, produced a dramatic but transient hyperemia (Figs. 7 a and b).

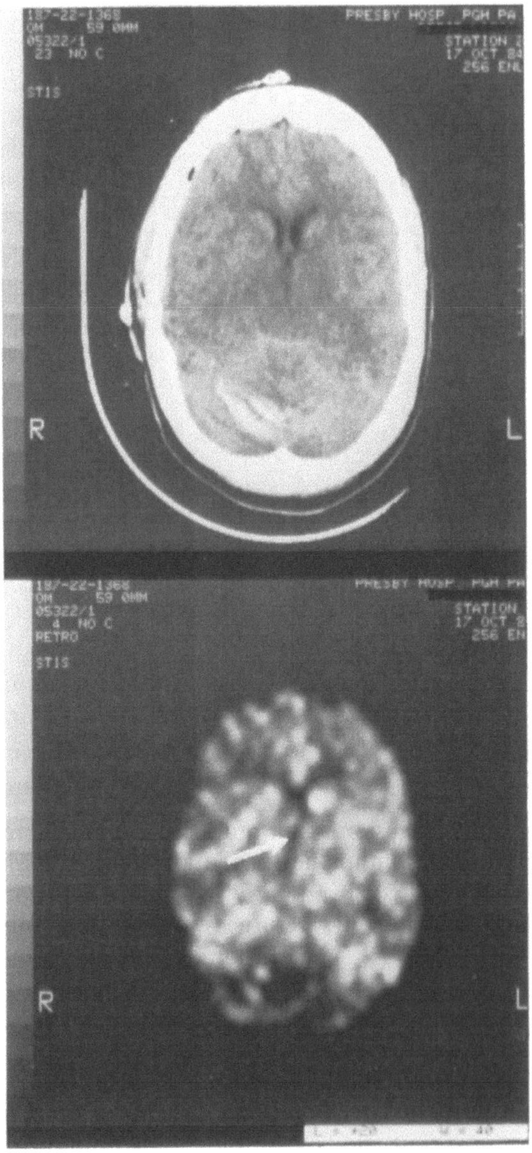

Fig. 8 a. Following clipping of a basilar tip aneurysm this women was paretic on her right side with dysconjugata gaze. While the CT scan disclosed no significant abnormality, the accompanying CBF study revealed a small region with absent perfusion within the right thalamus

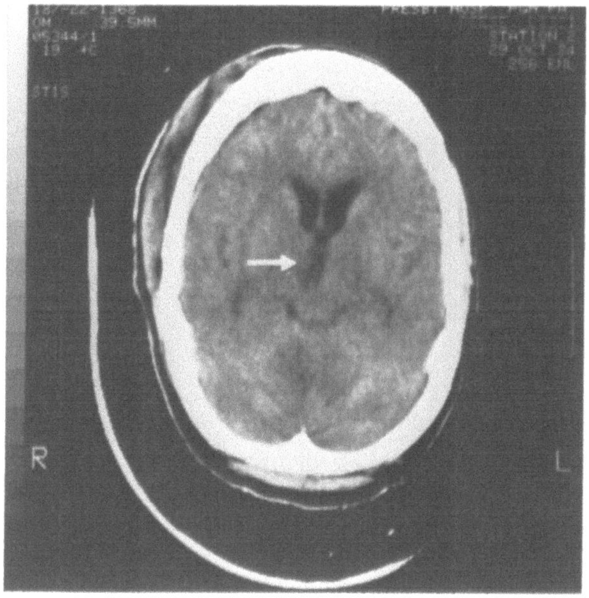

Fig. 8 b. A CT study obtained 12 days later demonstrated a CT pattern of infarction only in the region where flows had been below 10 cm^3/100 g/minute

Clinical Experience

The incorporation of the xenon/CT system within our standard scanning equipment has proven to be a useful diagnostic tool which adds an important new dimension to the information that CT imaging provides. Because of the relative ease of access to this technology, our system enables a flow study to be carried out in conjunction with a routine CT examination with only an additional 20 minutes required within the scanning facility. The transition from CT imaging to blood flow acquisition requires only an additional two to three minutes to connect two cables from the external delivery system and to fill the reservoir bag.

We have performed more than 1,800 flow examinations at the University of Pittsburgh on more than 1,500 patients, including 1,350 awake individuals who were at least partially ambulatory outpatients and 150 intubated patients with altered mental status. The CBF studies on awake patients have been well-tolerated by over 95% of the participants, resulting in satisfactory and useful clinical information from over 90% of this group. Motion artifact has been the major problem in the other 10%. In the intubated patients, 100% acquisition of useful information has been accomplished. In six patients in which intracranial pressure (ICP) was recorded during xenon inhalation, no consistent alteration of ICP during

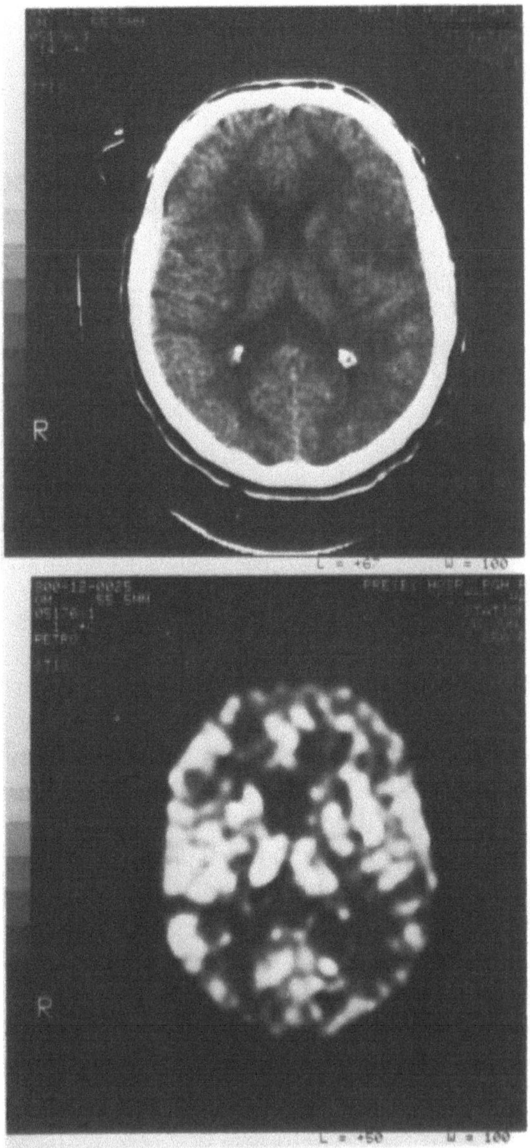

a

b

Fig. 9 a and b. This CT study (a) and accompanying CBF (b) examination were obtained 24 hours after this gentleman fell while jogging and sustained a left hemispheric stroke. Despite a high-grade carotid stenosis due to ICA dissection, the CT study revealed edema. The demonstration of hyperemia on the flow study suggests reperfusion of the ischemic region

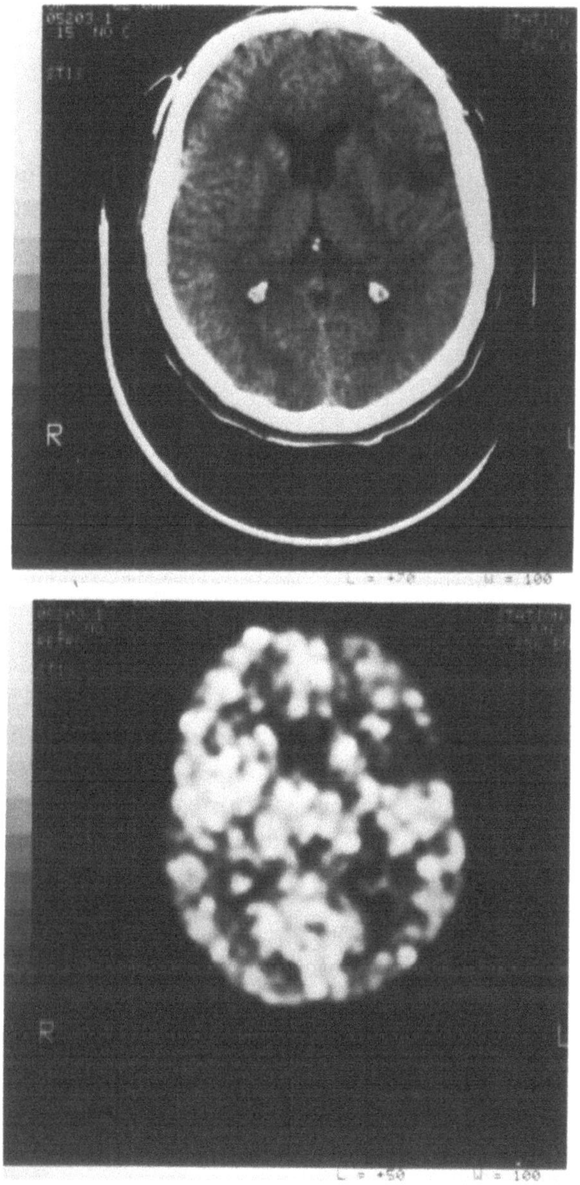

c

d

Figs. 9 c and d. The follow-up CT and CBF study three weeks later showed localized encephalomalacia and the absence of flow only in this region. Normal flow values within the remainder of the MCA distribution did not suggest the need for flow augmentation despite persistent high-grade ICA stenosis

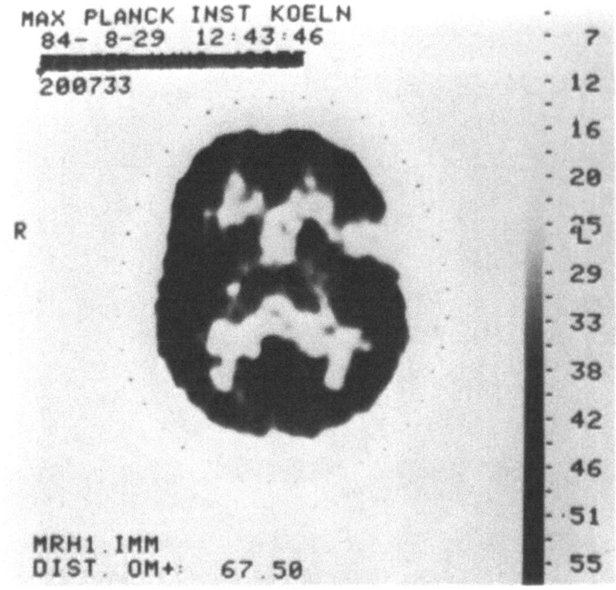

Fig. 9e. A positron emission tomographic study of glucose metabolism was obtained two months after the delayed CBF study (Fig. 9c). It demonstrates a high degree of correlation between flow and metabolism

xenon inhalation was found. In each of these cases, however, the patient was either moderately or severely hyperventilated throughout the examination[49].

While computerized tomography has proven useful for the triage of "stroke" patients, those with cerebral infarctions do not begin to show CT alterations of mass effect and density until at least 6–12 hours following a severe ischemic insult. The incorporation of xenon blood flow measurements has made CT technology sensitive to the earliest flow events that accompany cerebral infarction[7, 23, 51]. With xenon/CT, both small basal ganglion infarctions (Figs. 8a and b) and cortical lesions with absent or severely reduced flow have been readily evident and frequently seen within the first six hours after the onset of deficit. The absence of measurable CBF ($< 3\,cm^3/100\,g$/minute) has consistently been followed by a delayed CT appearance of infarction. Luxury perfusion with reperfusion has also been evident (Figs. 9a–e). An extensive laboratory study of xenon/CT low flow thresholds for EEG alteration and subsequent infarction is currently being conducted.

Because of our interest in chronic occlusive vascular disease, more than half of the patients we have studied have had single or multiple vessel occlusive disease. For these individuals, no effort was made to interpret residual flow reserve capacity earlier than four weeks following an

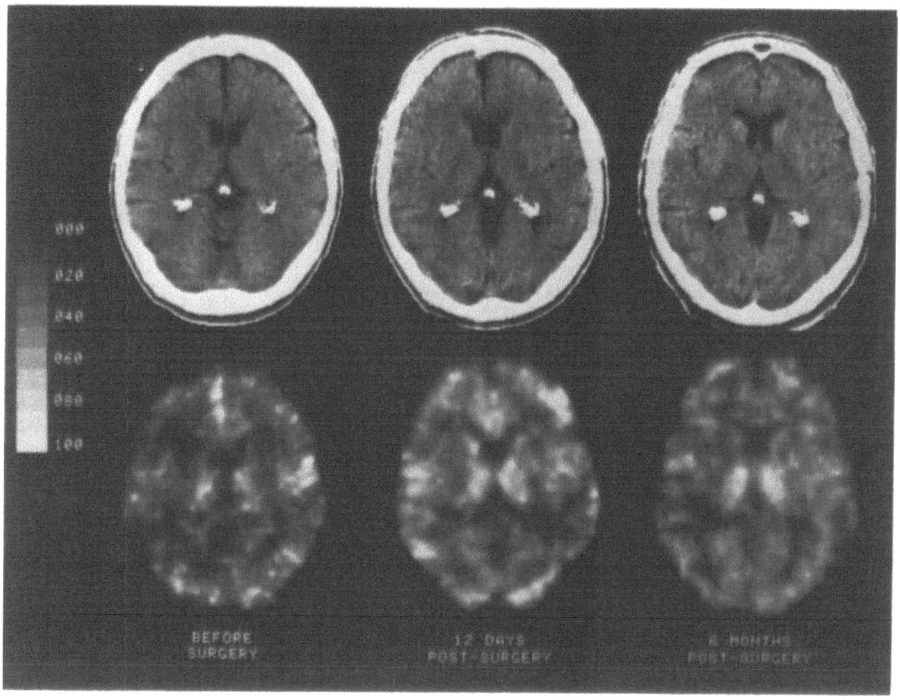

Fig. 10. This 78-year-old gentleman developed daily episodes of vertigo, unsteadiness of gait and confusion when standing. He was found to have an occlusion of his right common, internal and external carotid arteries with a high grade siphon stenosis of his left carotid artery. The right vertebral artery was occluded and the left highly stenotic. A preoperative global low flow pattern was corrected following a left STA/MCA bypass. He has been asymptomatic for two years following surgery. (Permission, Yonas, Ref. 44)

infarction. This delay allowed time for the recoupling of flow and metabolism to occur and for encephalomalacia to become apparent on the CT image. In the vast majority of these patients, and despite often worrisome angiographic patterns, we found no evidence of flow compromise within remaining tissues even in cases of significant adjacent atrophy. The higher resolution of xenon/CT CBF imaging does avoid the partial volume sampling error inherent in most other clinical CBF methodologies. It thereby more accurately reflects the adequacy of flow within remaining tissues in individuals with only cortical lesions.

In 12 patients, 11 of whom had multiple vessel occlusive disease, we were able to identify either a regional or global flow reduction within normal appearing brain tissue on CT[54, 56]. All but one of these individuals had experienced recurrent, positionally dependent, disabling TIAs, either hemispheric or nonhemispheric in nature despite anticoagulation. Flow

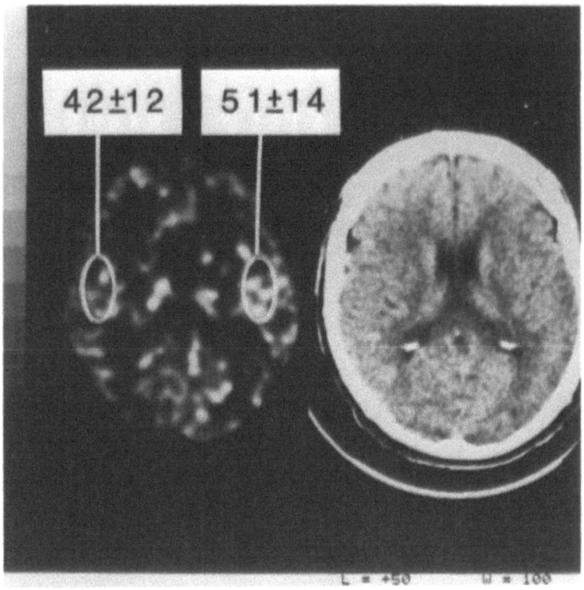

Fig. 11 a. This 58-year-old man had sustained a right hemispheric stroke two months earlier probably at the time of ipsilateral ICA occlusion. The CT scan suggests a small internal capsule lesion on the right side and the accompanying CBF study shows a moderate reduction of right hemispheric flow values compared to the left side

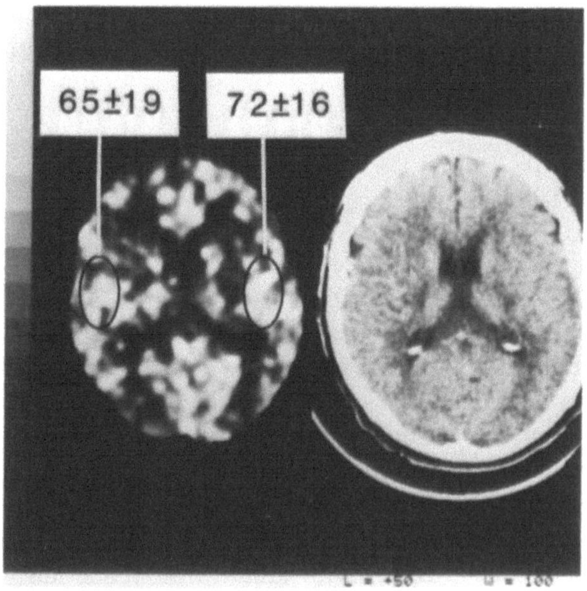

Fig. 11 b. Thirty minutes after the intravenous introduction of one gram of diamox, a 50% increase of CBF values is apparent within all vascular territories implying that there was no compromise of flow reserves. The study instead suggests that deafferentation with reduced metabolism and appropriately reduced CBF was present on the right side secondary to the more central brain injury

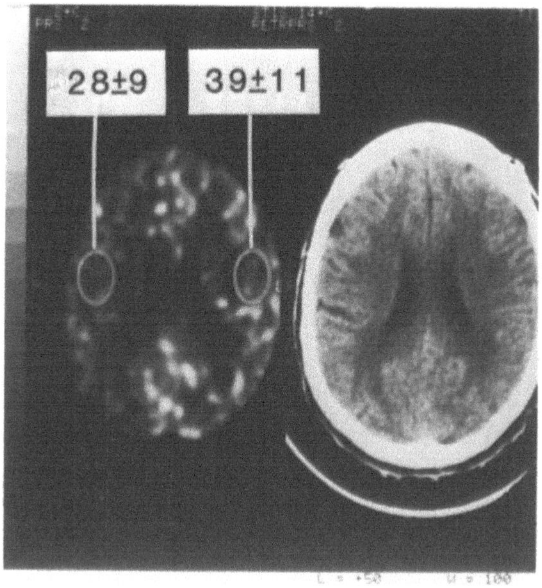

Fig. 12 a. This 63-year-old man had frequently recurrent episodes of left hand weakness and numbness which occurred while sitting or standing. The baseline CBF study demonstrated reduced flows within the right hemisphere despite absence of a CT-defined abnormality

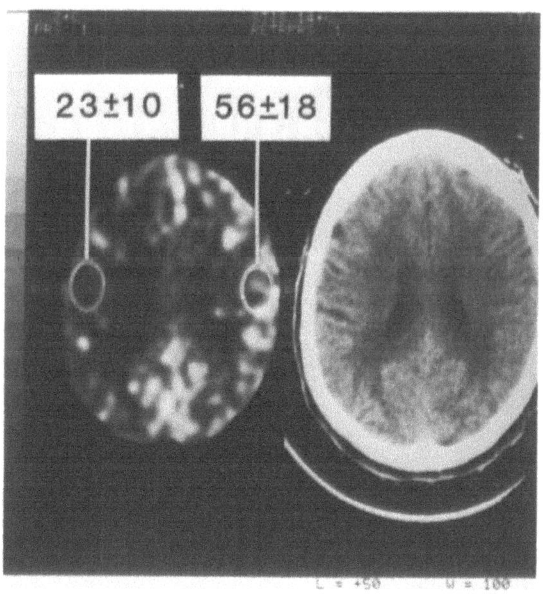

Fig. 12 b. Thirty minutes after the IV administration of diamox (one gram), his end-tidal pCO_2 had fallen from 36 to 32 mm Hg. The accompanying CBF study, however, shows a dramatic elevation of flow values in all regions except within the right MCA territory where no flow increase is observed. This study demonstrates a compromised flow reserve within the right MCA territory

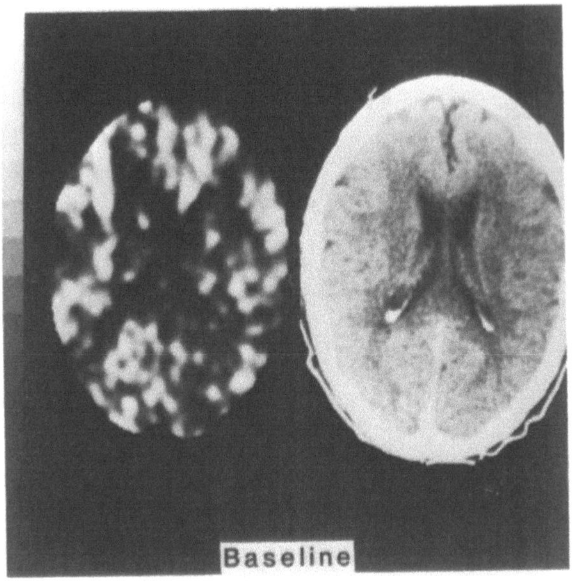

Fig. 13 a. This 42-year-old woman was found to have a malignant tumor of the right base of the skull for which occlusion of the right ICA was contemplated as possibly being necessary to accomplish a complete tumor removal. She initially displayed no neurological deficit despite a 15-minute trial ICA occlusion. Her initial flow study was normal prior to a second trial balloon occlusion of the right ICA

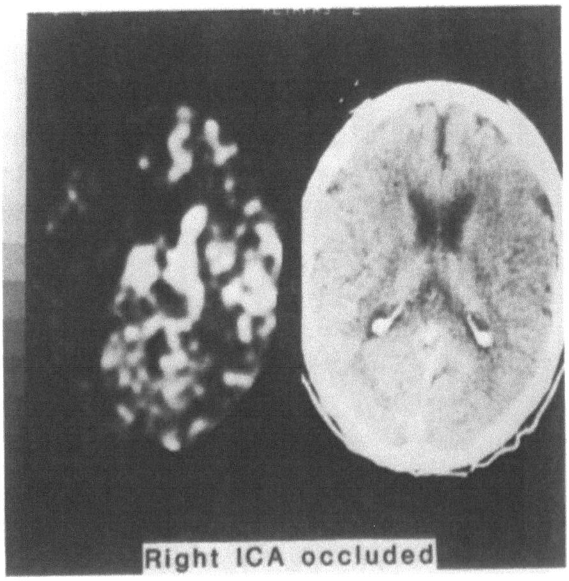

Fig. 13 b. Her repeat study was performed 20 minutes later, five minutes after balloon occlusion, with still no evidence of neurological dysfunction. The CBF study however showed a significant reduction of flow within the right anterior and middle cerebral distributions with mean flow values in the right MCA territory in the 18–25 cm^3/100g/minute range. Her pCO_2 (34 mm Hg) and blood pressure (130/80 mm Hg) were the same during both examinations. The study suggests an increased risk associated with permanent carotid occlusion despite the absence of apparent CNS dysfunction

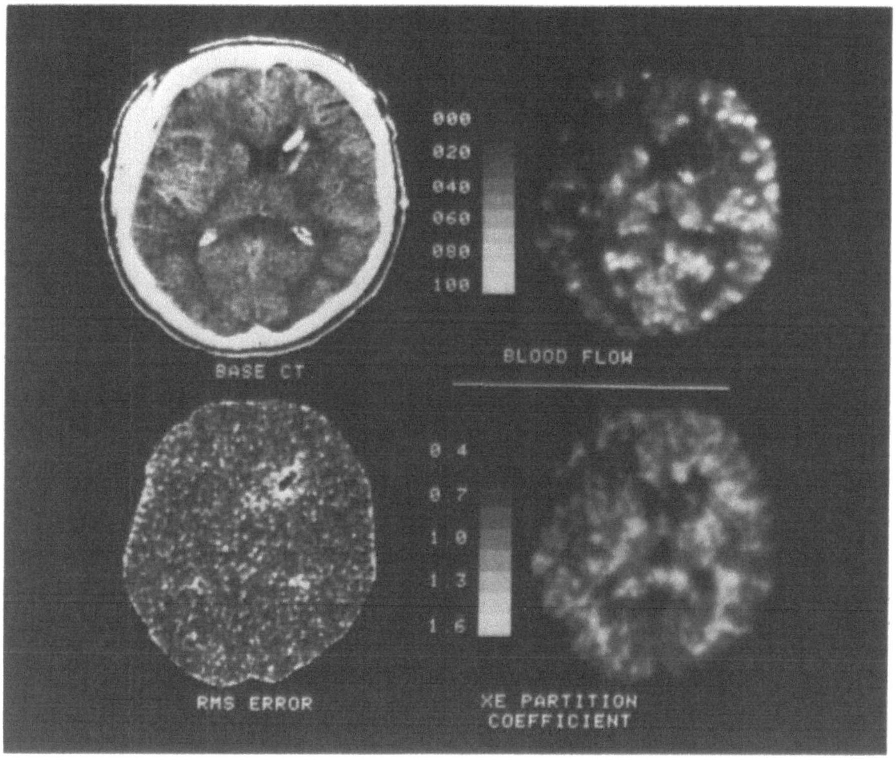

Fig. 14. Matching CT/CBF studies, partition coefficient, and confidence image are displayed in a patient with a right middle cerebral aneurysm. The delayed onset of left hemiparesis is presumably due to vasospasm. Reduced right MCA flow values in the 20–40 cm^3/100 g/minute range are evident (upper right). This is a good quality study with CT artifact (RMS error) evident only near the ventricular catheter. The normal reversal of flow versus lambda (lower right) is evident with deep white matter retaining high lambda values (> 1.0) and low flows (< 30 cm^3/100 g/minute)

augmentation procedures, such as internal or external carotid endarterectomy, vertebral reconstruction, or EC/IC bypass, proved beneficial in stopping recurrent disabling symptomatology in all of these individuals and resulted in a prolonged global elevation of CBF values (Fig. 10). A global redistribution and elevation of CBF has also been reported by Meyer following bypass surgery[31].

Because brain function and the accompanying CBF can be reduced by even small lesions within the basal ganglia, static blood flow information alone, if values are low, may not reflect the status of flow reserves. For this reason, we have begun using acetazolamide (1 gram) as a stress test of flow reserves[45]. In 40 patients this flow challenge has produced a dramatic flow elevation from the baseline study 30 minutes earlier ranging from 25–60%

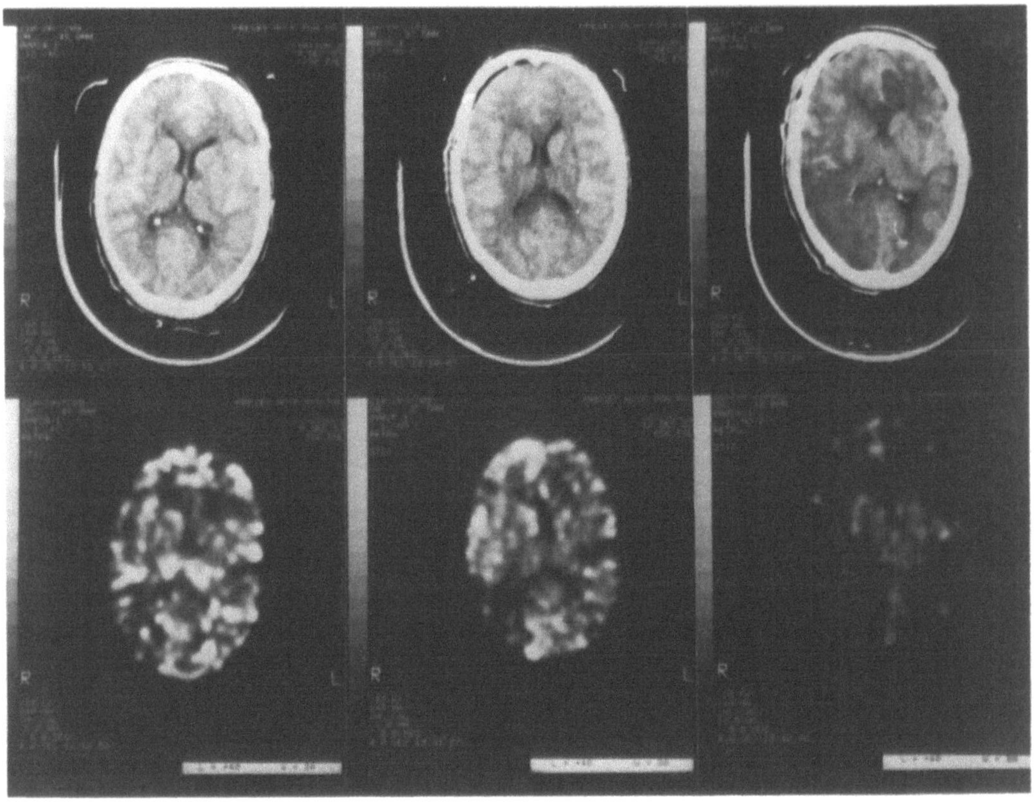

Fig. 15. This series of CT/CBF studies was obtained in a 50-year-old woman who survived subarachnoid hemorrhage (SAH) due to an anterior communicating aneurysm. The initial study (CT above left and flow study below left) was obtained five days post SAH when the patient was neurologically intact. Note reduced right posterior parietal flows. The second study (center CT above and CBF map below) was obtained the evening following surgical treatment of the aneurysm when the patient was slow to awaken. Note hyperemia in area of surgery and markedly reduced flows in right posterior parietal region. The third study was obtained the day following surgery when the patient was comatose. Note the transition of the lowest flow regions on the prior study to CT-defined infarctions. Also note the near absence of any cortical CBF. Patient progressed to brain death. The CBF gray scale was 0–80 cm^3/100 g/min for these studies

in most individuals with a regionally dampened, absent, or even paradoxical lowering presumably due to a "steal" phenomenon in reserve compromised patients (Figs. 11 a, b, and 12 a, b).

Conversely, we found dramatic decreases of CBF within a group of 20 patients that we evaluated by clinical and Xe/CT and CBF criteria for tolerance to carotid occlusion. While in most individuals the acute trial balloon-occlusion of one ICA produces either no or a slight reduction of CBF in all regions, two individuals demonstrated a severe reduction (mean flows 20–30 cm^3/100 g/minute) of primarily ipsilateral but also global flow without definable neurological deficit (Figs. 13a and b). Our initial experience suggests that these latter individuals are at risk from permanent ICA occlusion and we suspect if the ICA must be occluded in such a case, that they warrant either careful attention to maintain perfusion pressures and/or a prophylactic procedure to augment regional reserve flows[10].

Patients with delayed neurological deficits due presumably to vasospasm following aneurysmal rupture have also demonstrated dramatic alterations of CBF that have been either reversible or progressive in nature (Figs. 14 and 15)[52]. This type of CBF information has been helpful in guiding the treatment of vasospasm as well as helping select the least morbid time for aneurysm surgery.

In patients with focal or generalized mass effects due to head trauma, xenon/CT CBF information has demonstrated both focal and remote effects on CBF (Figs. 16a and b)[49]. The multicompartmental nature of the cranium is made evident since each dural compartment may be affected differently. The manipulation of pCO_2 has also been shown to produce variable alterations of CBF within different brain regions depending on the location and extent of injuries (Figs. 17a–c)[57]. We have found in head-injured patients that the CBF response to CO_2 manipulation, either higher or lower, also serves as an important guide for the extent and duration of hyperventilation therapy.

Other Clinical Uses

We and others have used Xe/CT to examine patients with tumors, seizures, dementia, coma, and brain death. Differential patterns of CBF and lambda have been noted between high and low grade gliomas as well as with metastatic lesions[33]. In a series of patients with intractable epilepsy, xenon/CT imaging has been useful in localizing seizure foci, with findings similar to those found with PET scanning. Generally, we have seen severely reduced cortical flows in patients with advanced dementias and in coma (flows ranging from 20–30 cm^3/100 g/minute). As part of a center project to study dementias in aging, we are also using the Xe/CT procedure to examine the effects of Alzheimer's disease on CBF. In midwork, we have seen a change in blood flow patterns in presymptomatic Alzheimer's disease, with reduced cortical flow out of proportion with cerebral atrophy in some areas while other regions (primariliy in phylogenetically older structures) exhibit irregularly increased flow patterns. In patients in whom a diagnosis of brain death is being questioned, the absence of measurable blood flow in the mid and upper brain stem, cerebellum and hemispheres has been shown by

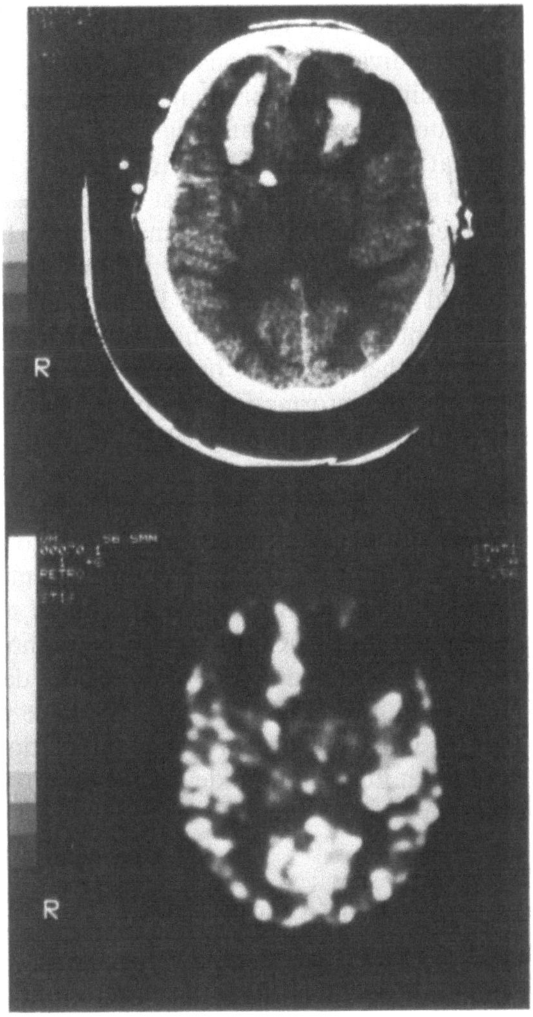

Fig. 16 a. The initial CT and CBF study was obtained four days post-trauma when the patient developed decerebrate posturing and dilated pupils, despite a high normal and stable ICP. The CBF study discloses very low basal ganglia flows accompanying the severely reduced left frontal flows

Xe/CT to correlate consistently with other standard, and often more invasive, examinations used to certify brain death (Fig. 18)[2, 3].

Limitations and Problem Areas

The anesthetic properties of xenon, its associated effect on blood flow, and the accompanying radiation exposure all warrant discussion.

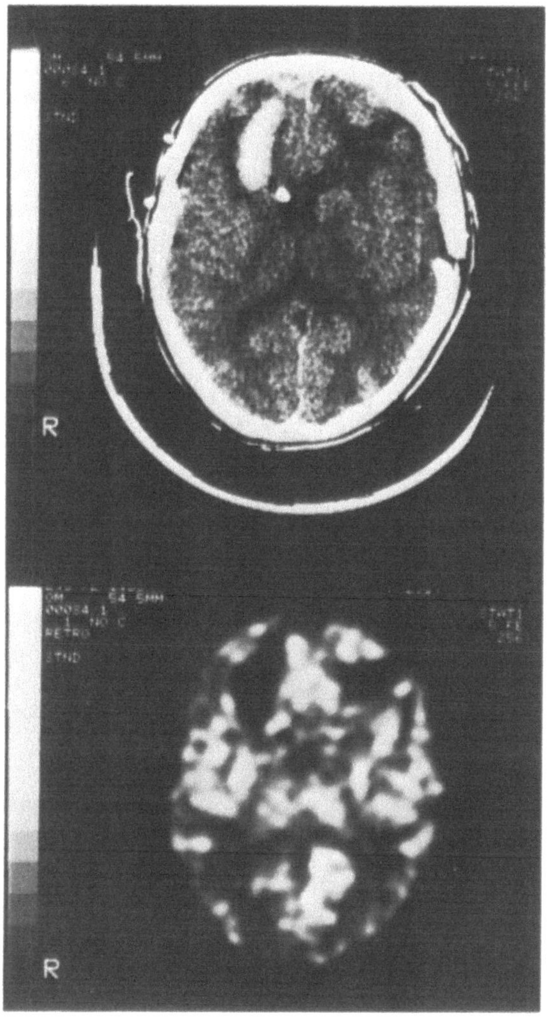

Fig. 16 b. A return to more normal central flow values is apparent following removal of the left frontal hematoma. (Permission, Wozney, Ref. 39)

While the incidence of intolerance to inhalation of xenon appears to be low at a concentration of about 30%[28], individual intolerances to this concentration still exist. Some people are very sensitive to xenon and show either an early anesthetic state effect or agitation often manifested as laughter or a brief unresponsiveness. In a review of 1,750 studies performed at five institutions, reports of significant side-effects included respiratory slowing in 3%, seizures in 0.2%, delayed headache in 0.4% and nausea or vomiting in 0.2%[28]. Individuals who are most sensitive to xenon seem to develop an earlier slowing and irregularity of respiratory pattern; we now

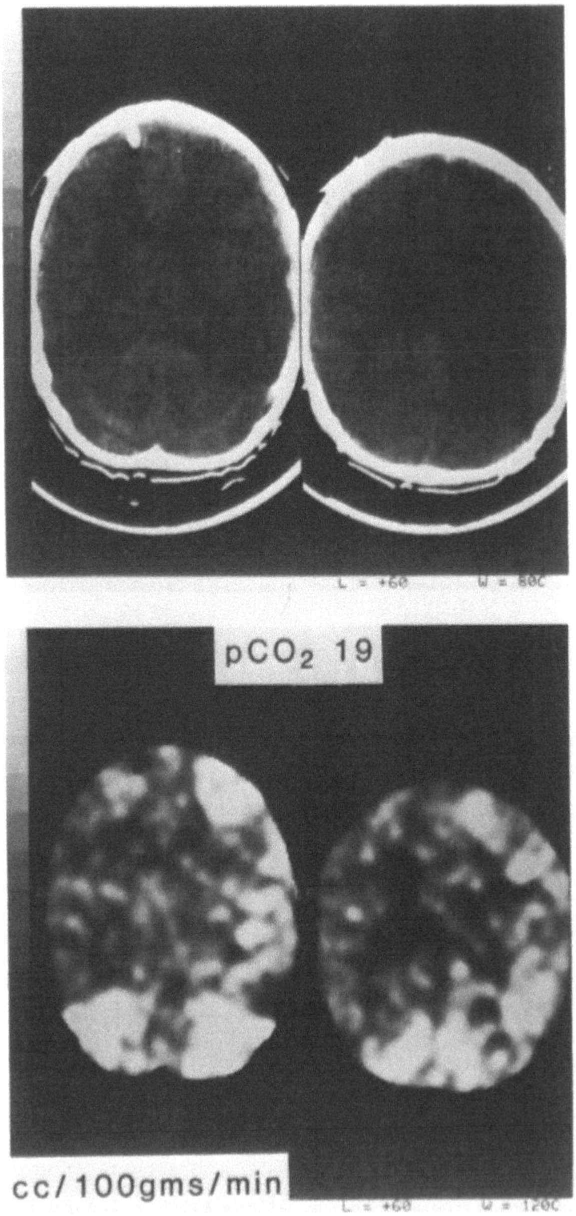

a

b

Figs. 17 a–c. This three-year-old had remained comatose four days following a closed-head injury with the ICP elevated beween 25–30 mm Hg despite medical therapy. The CT-imaged (Fig. 17 a) and accompanying blood flow studies at two brain levels (Fig. 17 b and c) are displayed at pCO_2 of 19 and 30 mm Hg obtained by ventilator manipulation. The gray scale is 0–120 cm^3/100 g/min. Note that, despite CO_2 manipulation, flow values within the left hemisphere and cerebellum remained elevated with a focal shift to even higher flows with CO_2 lowering. The lowering of pCO_2 to 19 mm Hg also caused a dramatic decrease of CBF to near ischemic levels in the right hemisphere. No lowering of ICP was found with pCO_2 lowering presumably due to the paradoxical increase of focal CBF within the left hemisphere

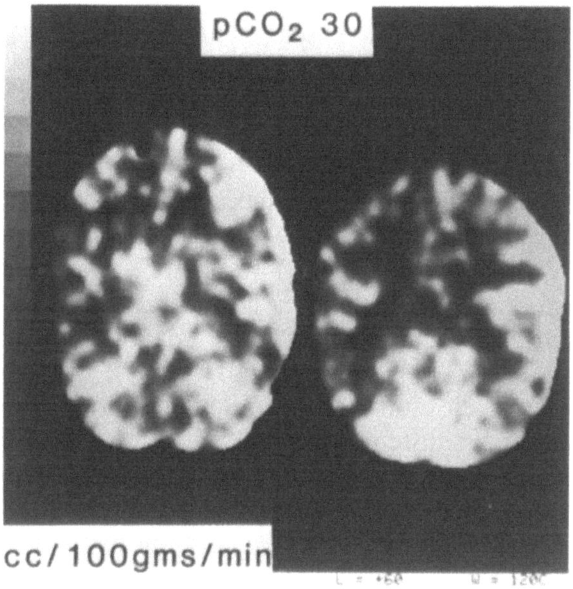

c

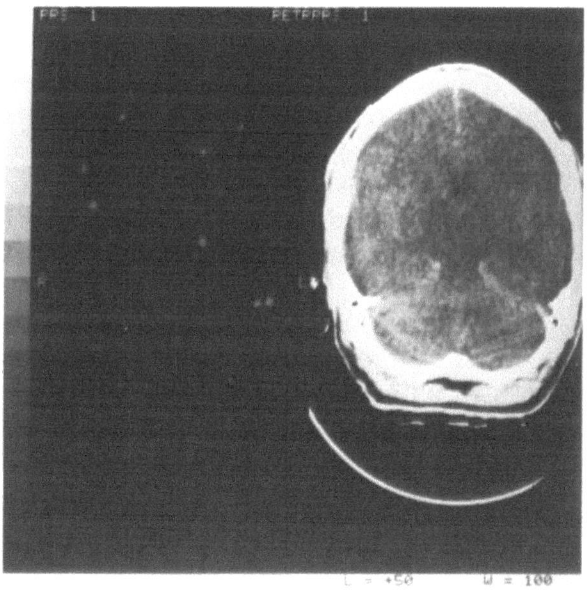

Fig. 18. This 32-year-old man was found one hour after a stab wound to the chest with a systolic blood pressure of 20 mm Hg. Despite a heroic thoracic surgical procedure, he remained without evidence of neurological function. On day two postinjury, while hypothermic and despite a blood pressure of 120/80, he retained essentially no CBF within any brain region

use this as an indication that a study should be terminated. Although bradypnea, often accompanied by a significant decrease of pCO_2, occasionally occured in our earlier studies when the gas was inhaled beyond a five-minute period, it is now rarely encountered with shorter studies (< 4.5 minutes) in which 0.8% CO_2 has been added to the gas mixture[1].

The effect of xenon on blood flow has been evaluated by a number of investigators[19, 24, 34, 35]. Although one report suggests a significant differential elevation of only neocortical flows in rats[24], this extent or pattern of flow augmentation has not been observed by others. Radiolabeled microsphere studies in nonhuman primates have demonstrated either no flow alteration[35] or up to a 17% homogenous increase of flow in all brain regions[20]. An examination of CBF alterations in normal volunteers using Xe-133 has identified a more highly variable degree of flow enhancement, averaging 30% during a 4–5 minute period of xenon inhalation[34]. Because the flow values actually derived from the stable xenon/CT methodology do not appear to be as severely divergent from established norms of flow[7, 29, 39] as suggested by some of these studies, the true significance of these measurements is yet to be defined. Our impression after examining more than 2,000 CBF procedures in human and animal studies is that an increase of flow, to whatever extent it may influence the flow calculation, is minimal and may be clinically useful, serving as a challenge of flow reserves similar to that obtained by inhaling added pCO_2. In addition, the elevation of CBF during xenon inhalation only serves to elevate measured flow values so that low or absent flow values should be of greater clinical relevance. The tendency of xenon to elevate CBF is of concern in patients with elevated intracranial pressure. While our initial observations demonstrated no ICP rise during xenon inhalation, a larger study has shown that a significant elevation of ICP can occur in head-injured patients. Careful ICP monitoring will therefore be required until the relationship of ICP and xenon inhalation is fully defined.

A severely limiting aspect of relatively high resolution flow imaging is that no motion can be tolerated during the 6 to 6–$^1/_2$ minutes that are required for acquiring baseline and enhanced scans. With a relatively high degree of resolution, significant misregistration can occur with movement as small as a few millimeters in any plane of motion. To aid in the maintenance of head position, an evacuable bean bag is used for support, and an extensive introduction emphasizing the importance of remaining still is given to each patient.

Radiation exposure is another aspect of the methodology which must be considered when undertaking a Xe/CT study. The dose of radiation per CT cut, depending on the exposure factors, can range from 2–3.5 rads. A flow study will deliver from 8–28 rads to each level studied, depending on the number of baseline and enhanced scans obtained. Although these are considered to be clinically acceptable levels comparable to other radiographic procedures such as cerebral angiography, it is not an insignificant dose of radiation. The radiation field, however, is highly collimated, and contiguous scans are avoided to minimize an overlapping band of even higher radiation. Exposure to the eye is avoided. Further, in comparison with other body organ systems, the brain is considered less radiosensitive so that the risk associated with even the highest possible exposure delivered by a single or multiple xenon/CT CBF study is relatively low[37].

Conclusion

Xe/CT flow information appears to be clinically useful in the diagnosis and management of a broad spectrum of clinical disorders. It also appears to be a useful tool for the experimental study of the cerebral circulation with recent work being extended to the study of solid abdominal organs[18].

We have found xenon/CT CBF mapping to be a useful new blood flow methodology and believe that, as CT technology improves, we will be able to obtain still better flow information with less accompanying radiation exposure and/or a reduction in the level of xenon inhalation required.

Acknowledgment

I would like to extend a special thank you to Gail Schwartzmiller and Cynthia Gill for preparation of and editorial assistance with the manuscript.

References

1. Clark F, Yonas H, Gur D: Alteration of respiration rate and end exhalation CO_2 during inhalation of subanesthetic levels of xenon gas. J Comput Assist Tomogr (in press)
2. Darby JD, Yonas H, Brenner R: Brainstem death with persistent EEG activity: evaluation by xenon-enhanced computed tomography. Critical Care Medicine 15 (5): 519–521.
3. Darby JD, Yonas H, Gur D, Latchaw RE: Xenon-enhanced computed tomography in brain death. Archives of Neurology 44: 551–554.
4. Dhawan V, Goldiner P, Ray C, Conti J, Rottenberg DA (1982) Mass spectrometric measurements of end-tidal xenon concentration for clinical stable xenon/computerized tomography cerebral blood flow studies. Biomed Mass Spectrom 9 (6): 241–245
5. Dhawan V, Haughton VM, Thaler HT, Lu HC, Rottenberg DA (1984) Accuracy of stable xenon/CT measurements of regional cerebral blood flow: effect of extrapolated estimates of brain-blood partition coefficients. J Comput Assist Tomogr 8 (2): 208–212
6. Drayer BP, Wolfson SK Jr, Reinmuth OM, *et al* (1978) Xenon enhanced computed tomography for the analysis of cerebral integrity, perfusion, and blood flow. Stroke 9: 123–130

7. Drayer BP, Gur D, Yonas H, Wolfson SK Jr, Cook EE (1980) Abnormality of the xenon brain-blood partition coefficient and blood flow cerebral infarction: an in vivo assessment using transmission computed tomography. Radiology 135 (2): 349–354
8. Eidelman BH, Hospodar M, Ratcliff G, Gur D, Yonas H, Latchaw R, Deeb Z (1985) Dementia and stable xenon/CT cerebral blood flow in multiple sclerosis. J CBF and Metabol 5 (1): s45–46
9. Eidelman BH, Hospodar M, Gur D, Yonas H, Latchaw R, Deeb Z (1985) A comparison of cerebral blood flow and NMR imaging in chronic multiple sclerosis. J CBF and Metabol 5 (1): s47–48
10. Erba M, Yonas H, Latchaw RE: Balloon occlusion monitored by xe/CT CBF (in preparation)
11. Good WF, Gur D, Shabason L, Wolfson SK Jr, Yonas H, Latchaw RE, Herbert DL, Kennedy WH (1982) Errors associated with single-scan determinations of regional cerebral blood flow by xenon-enhanced CT. Physics in Medicine and Biology (London) 27 (4): 531–537
12. Good WF, Gur D, Yonas H, Herron JM: Errors in CBF determinations by xenon-enhanced CT due to estimation of arterial xenon concentration (submitted)
13. Gur D, Yonas H, Wolfson SK Jr, Herbert D, Kennedy W, Drayer B, Shabason L (1981) Xenon and iodine enhanced cerebral CT: A closer look. Stroke 12 (5): 573–578
14. Gur D, Yonas H, Herbert DL, Wolfson SK, Kennedy WH, Drayer BP, Gray J (1981) Xenon enhanced dynamic computed tomography: multilevel cerebral blood flow studies. J Comput Assist Tomogr 5 (3): 334–340
15. Gur D, Wolfson SK, Yonas H, Good W, *et al* (1982) Progress in cerebrovascular disease: LCBF by xenon-enhanced CT. Stroke 13: 750–758
16. Gur D, Good WF, Wolfson SK Jr, Yonas H, Shabason L (1982) In vivo mapping of local cerebral blood flow by xenon-enhanced computed tomography. Science 5: 1267–1268
17. Gur D, Herron HM, Molter BS, Good BC, Albright RE, Miller JN, Drayer BP (1984) Simultaneous mass spectrometry and thermoconductivity measurements of end-tidal xenon concentrations: A comparison. Med Phys 11 (2): 208–212
18. Gur D, Good WF, Herbert DL, Yonas H, Wozney P, Thiel DH van, Wolfson SK Jr (1985) Mapping of human blood flow in the liver by the xenon/CT method. JCAT 9 (3): 447–450
19. Gur D, Yonas H, Jackson DL, Wolfson SK Jr, Rockette H, Good WF, Cook EE, Arena VC, Willy JA, Maitz GS (1985) Simultaneous measurements of cerebral blood flow by the xenon/CT method and the microsphere method: A comparison. Invest Radiol 20: 672–677
20. Gur D, Yonas H, Jackson DL, Wolfson SK Jr, Rockette H, Good WF, Maitz GS, Cook EE, Arena VC (1985) Measurements of cerebral blood flow during xenon inhalation as measured by the microsphere method. Stroke 16: 871–874
21. Haughton VM, Donesan JF, Walsh PR, Syvertsen A, Williams A (1980) A clinical evaluation of xenon enhancement for computed tomography. Invest Radiol 15 [6 Suppl]: s160–163

22. Harrington TR, Manwaring K, Hodak J (1986) Local basal ganglia and brain stem blood flow in the head injured patient using stable xenon-enhanced CT scanning. In: Miller JD, Teasdale GM, Rowan JV, Galbraith SL, Mendelow AD (eds) Intracranial pressure. Springer, Berlin Heidelberg New York Tokyo
23. Hughes R, Yonas H, Gur D, *et al*: Stable xenon-enhanced CT in the first eight hours of cerebral infarction (in preparation)
24. Junck L, Dhawan V, Thaler HT, Rottenberg DA (1985) Effects of xenon and krypton on regional cerebral blood flow in the rat. J CBF and Metabol 5: 126–132
25. Kelcz F, Hilal SK, Hartwell P, Joseph PM (1978) Computed tomographic measurement of the xenon brain-blood partition coefficient and implication for regional cerebral blood flow: a preliminary report. Radiology 127: 385–392
26. Kishore PR, Rao Gu, Fernandez RE, Keenen RL, Arora GD, Gadisseux P, Steward LM, Wist AD, Fatouros PP, Dillard D, *et al* (1984) Regional cerebral blood flow measurements using stable xenon enhanced computed tomography: A theoretical and experimental evaluation. J Comput Assist Tomogr 8 (4): 619–630
27. Koeze TH, Wylie IH (1980) Determination of cerebral blood flow with the EMI CT scanner. Br J Radiol 53 (629): 410–415
28. Latchaw RE, Yonas H, Pentheny SL, Gur D: Adverse reactions to the inhalation of 32% stable xenon for Xe/CT cerebral blood flow determination. Radiology (in press)
29. Meyer JS, Hayman LH, Yamamoto M, Sakal F, Nakajima S (1980) Local cerebral blood flow measured by CT after stable xenon inhalation. AJR 135 (2): 239–251
30. Meyer JS, Yamamoto M, Hayman LA, Sakai F, Nakajima S, Amstrong D (1980) Cerebral embolism: local CBF and edema measured by CT scanning and xenon inhalation. Neurol Res 2 (2): 101–126
31. Meyer JS, Nakajima S, Dkabe T, Amano T, Centeno R, Len YY, Levine J, Levinthal R, Rose J (1982) Redistribution of cerebral blood flow following STA/MCA bypass in patients with hemispheric ischemia. Stroke 13 (6): 774–784
32. Meyer JS, Kitasawa Y, Tanahashi N, Tachibana H, Kanoula P, Cech DA, Rose JE, Grossman RG (1985) Pathogenesis of normal-pressure hydrocephalus—preliminary observations. Surg Neurol 23 (2): 121–133
33. Nakamura D, Nomura K, Segawa H, Takakura K, Nakasome T, Yoshimasu N, Ueda Y, Kimura K, Nasai M (1985) Cerebral blood flow measured by xenon-enhanced CT in brain tumors. No Shinkei Geka 13 (1): 37–43
34. Obrist WD, Jaggi JL, Harel D, Smith DS (1985) Effect of stable xenon inhalation on human CBF. J CBF and Metabol 5 (1): s557–558
35. Panos PP, Fatouros R, Kishore PRS, Hall JA Jr, West AO, Dewitt DS, Marmarou A, Keenan R, Kantos H (1985) Comparison of improved stable xenon/CT method for cerebral blood flow measurements with radiolabeled microspheres technique. Radiology 158 (1): 334
36. Pittinger CB, Moyers J, Kuhlen SC (1953) Clinical pathologic studies associated with xenon anesthesia. Anesthesiology 14: 10

37. Report of the National Institutes of Health Ad Hoc Working Group to Develop Radioepidemiological tables. Office of the Director of National Institutes of Health (Publ). No. 85–2748. January, 1985
38. Rottenberg DA, Lu HC, Kearfott KJ (1982) The in vivo autoradiographic measurement of regional cerebral blood flow using stable xenon and computerized tomography: The effect of tissue heterogeneity and computerized tomography noise. J CBF and Metabol 2 (2): 173–178
39. Segawa H (1985) Tomographic cerebral blood flow measurement using xenon inhalation and serial CT scanning: Normal values and its validity. Neurosurg Rev 8 (1): 27–33
40. Sekhar LN, Schramm VL Jr, Jones NF, Yonas H, Horton J, Latchaw RE, Curtin H (1986) Operative exposure and management of the petrous and upper cervical internal carotid artery. Neurosurgery 19: 967–982
41. Thaler HT, Baglivo JA, Lu HC, Rottenberg DA (1982): Repeated least squares analysis of simulated xenon computed tomographic measurements of regional cerebral blood flow. J CBF and Metab 2 (4): 408–414
42. Tachibana H, Meyer JS, Rose JE, Kandula P (1984) Local cerebral blood flow and partition coefficients measured in cerebral astrocytomas of different grades of malignancy. Surg Neurol 21 (2): 125–131
43. Tachibana H, Meyer JS, Okayasu H, Shaw TG, Kandula P, Rogers RL (1984) Xenon contrast CT-CBF scanning of the brain differentiates normal age-related changes from multi-infarct dementia and senile dementia of Alzheimer type. J Gerontol 39 (4): 415–423
44. Tachibana H, Meyer J, Kitagawa Y, Tanahashi N, Kandula P, Rogers RL (1985) Xenon contrast CT-CBF measurements in parkinsonism and normal aging. J Am Geriatr Soc 33 (6): 413–421
45. Vorstrup S, Engell HC, Lindewald H, Lassen NA (1984) Hemodynamically significant stenosis of the internal carotid artery treated with endarterectomy. J Neurosurg 60: 1070–1075
46. Winkler S, Sackett J, Holden J, *et al* (1977) Xenon inhalation as an adjunct to computerized tomography of the brain. Preliminary study. Invest Radiol 12: 15
47. Winkler S, Turski P (1985) Potential hazards of xenon inhalation. AJN 6: 974–975
48. Wolfson SK Jr, Gur D, Yonas H: Autoregulation recorded by xenon/CT CBF measurements (in preparation)
49. Wozney P, Yonas H, Latchaw RE, Gur D, Good W (1985) Central herniation revealed by focal decrease in blood flow without elevation in ICP: A case report. Neurosurg 17 (4): 641–644
50. Yonas H, Wolfson SK Jr, Dujovny M, Boehnke M, Cook E (1981) Selective lenticulostriate artery occlusion in the primate—A highly focal cerebral ischemic model. Stroke 12 (5): 567–572
51. Yonas H, Grundy B, Gur D, Shabason L, Wolfson SK Jr, Cook E (1981) Side effects of xenon inhalation. J Comput Assist Tomogr 5: 591-592
52. Yonas H, Wolfson SK Jr, Gur D, Latchaw RE, Good WF, Leanza R, Jackson DL, Jannetta PJ, Reinmuth OM (1984) Clinical experience with the use of xenon-enhanced CT blood flow mapping in cerebral vascular disease. Stroke 15 (3): 443–449

53. Yonas H, Good WF, Gur D, Wolfson SK Jr, Latchaw RE, Good BC, Leanza R, Miller SL (1984) Mapping cerebral blood flow by xenon-enhanced computed tomography: Clinical experience. Radiology 152 (2): 425–442
54. Yonas H, Gur D, Good BC, Latchaw RE, Wolfson SK Jr, Good WF, Maitz GS (1985) Stable xenon/CT blood flow mapping in the evaluation of patients before and after extracranial/intracranial bypass surgery. J Neurosurg 62: 324–333
55. Yonas H, Wolfson SK Jr, Cook EE, Gur D (1986) Selective lenticulostriate arterial occlusion: A reproducible model of primate focal cerebral ischemia. Surg Neurol 25: 545–552
56. Yonas H, Steed DL, Latchaw RE, Gur D, Peitzmann AB, Webster MW: Relief of non-hemispheric symptoms in low flow states by anterior circulation revascularization: A physiological approach. J Vasc Surg (in press)
57. Yonas H, Snyder JV, Gur D, Good WF, Latchaw RE, Wolfson SK Jr, Grenvik A, Good BC (1984) Local cerebral blood flow alterations (Xe-CT method) in an accident victim: A case report. J Comput Assist Tomogr 8 (5): 990–991

Physiological, Inflammatory and Neuropathic Pain

CLIFFORD J. WOOLF

Cerebral Functions Research Group, Department of Anatomy, University College London, London (U.K.)

Contents

Introduction

The confusion amongst clinicans and basic scientists as to what precisely pain is, is emblematic of the difficulties that confront all who attempt to elucidate its pathogenesis. A major contributor to this state of affairs would appear to have been Charles Sherrington. At the turn of the century, with great insight, he focussed attention away from pain itself onto the stimulus in the periphery. In attempting to link stimulus and response in a quantifiable way, he argued that while pain may be caused by a diversity of circumstances, it was commonly associated with peripheral injury. For those categories of stimuli that were of sufficient intensity to threaten, or to produce tissue injury, he coined the term "noxious". Noxious stimuli he argued, are detected by specialized high threshold sense organs, "nociceptors" and these uniquely signal to the central nervous system, the occurence of tissue damage (Sherrington 1906). The appeal of Sherrington's analysis lies in its simplicity, for it implies that once a noxious stimulus activates a set of nociceptors, a chain of circumstances (sensory processing within the central nervous system) is initiated that concludes with the perception of a sensory state that we call pain. Such a view has for long dominated the study of the neurobiology of pain (Perl 1985, Willis 1985).

Two major problems beset this classical approach that interprets pain as being the conscious appreciation of activity in a sensory systems that qualitatively resembles other sensory systems such as those that subserve visual, auditory or tactile sensations, with its own specific receptors and pathways. The first is that acute pain is not differentiated from chronic pain and the second is that pain as a physiological response is not differentiated from pain as a disease state. Pain is not a unitary sensory state, rather it is a unifying descriptor that is chosen by us to collectively describe a wide range of different types of sensory states all of which are colored by unpleasant, uncomfortable and disturbing connotations.

In certain circumstances pain appears to possess characteristics similar to those found in other somatic and special sensory systems; a discrete threshold and a stable stimulus response curve. However, these are only found in trained subjects using "experimental pain" that is transient in its effects and can be terminated at any stage by the subject. This is similar to the pains we commonly experience in everyday life as pinprick, excessive heat or cold and minor chemical irritation. Because the stimulus intensity can be quantified and repeated and because there is an ordered relationship between stimulus and response this type of pain has been the subject of over 90% of all experimental research in the subject both in man and in animals, where the end point is some simple form of motor behavior (*e.g.,* the tail flick).

Although we use the same word in our everyday vocabulary to describe "physiological" and "clinical" pain this must not be taken to mean that similar neural mechanism are necessarily responsible. Physiological pain or nociception is a sherringtonian sensation—the perception of or reaction to a noxious stimulus. Clinical pain—an ongoing, distressing and agonizing sensory state—arises following tissue injury in the case of inflammatory pain or as a consequence of a disturbance of the normal functioning of the nervous system in the case of neuropathic pain.

The distinction between the three different forms of pain, while of great importance is not absolute. An intense noxious stimulus giving rise initially to "physiological" pain may damage tissue initiating an inflammatory response causing acute inflammatory pain, a clinical pain. Chronic inflammation may, with time, induce alterations in the nervous system (either peripheral or central) changing inflammatory pain to neuropathic pain, such that removal of the inflammation does not necessarily remove the pain.

This chapter will review our current understanding of those somatosensory systems that appear to be responsible for the different types of pain, emphasizing recent advances and with a particular focus on primary afferent neurones and the spinal cord.

The Primary Afferent Neurone

Under normal, nonpathological circumstances two categories of primary afferent fibers signal the detection of noxious stimuli; the thin myelinated A-delta afferent and the unmyelinated C-afferent. These afferents innervate skin, subcutaneous tissue, joints, muscle and viscera (Willis 1985). Although not exclusively activated by intense tissue damaging stimuli, the majority of these afferents in rats, cats and primates have high thresholds and therefore fullfil Sherrington's definition of nociceptors. Most A-delta nociceptors are sensitive to mechanical stimuli only (Burgess and Perl 1973, Fitzgerald and Lynn 1977, Georgopoulos 1977). A feature of C nociceptors, apart from their slow conduction velocity, is their polymodal sensitivity. These nociceptors respond to mechanical, chemical and thermal stimuli (Bessou and Perl 1969, Iggo and Ogawa 1971, La Motte *et al.* 1983). The chemosensitivity of C afferents is of considerable importance because it is this property that is likely to signal the liberation of chemicals from; injured cells, inflammatory cells, afferent terminals and from sympathetic efferents.

Early psychophysical data pointed to noxious stimuli evoking two pains, an early sharp pain and a second burning pain. The latency of these responses were compatible with the first being an A-delta fiber mediated phenomenon and the second being produced by C fibers (Lewis 1942, Cambell and La Motte 1983). The technique of microneuronography has, in recent years, transformed our analysis of nociceptor function in humans. Insertion of an electrode into nerve fascicles permits the direct recording of activity in single afferent fibers, whose conduction velocity can be measured and whose peripheral receptive field can be identified (Gybels *et al.* 1979, Torebjörk *et al.* 1984). The combination of this technique with controlled stimuli and a psychophysical analysis of the perceived sensory response is indeed very powerful and has confirmed that, as in other vertebrates, primary afferents in man are organized into specific subtypes with A-delta and C afferent fibers mediating nociception. The data from these studies appears at first to support the notion that the somatosensory system is organized as a series of "labeled lines", each of which signals a discrete sensory event with a particular temporal and modality characteristic (Ochoa *et al.* , 1985). Further support for this analysis is the claim that by using the microneurographic electrode to stimulate single, or small groups of afferents, it is possible to evoke discrete and distinct sensations (Konietzny *et al.* 1981, Torebjörk and Ochoa 1983). However, that the somatosensory system is organized as a relay of specific parallel sensory pathways has been seriously challenged on both technical and theoretical grounds (Wall and McMahon 1984). Normal cutaneous stimuli produce sensations by activating hundreds of afferents belonging to many different classes with different thesholds, temporal properties and modality specificities and it is this pattern of activity across the population, that contains the signal about the stimulus. Mechanical stimuli have been shown, for example, to produce a higher firing rate in polymodal C afferents than thermal stimuli although they are perceived to produce less pain (Gybels *et al.* 1979). These two viewpoints are not necessarily incompatible with each other in the sense that while cutaneous afferents clearly do have specific properties which signal a specific input to the central nervous system, the input in a particular "labeled line" must, except in highly artifical circumstances, be blended centrally with input in many other lines and it is this population signal that contains the totality of information concerning the location, onset, duration, magnitude and quality of a particular stimulus, noxious or otherwise. Equally important is the fact that afferent fibers trigger inhibitory as well as excitatory events centrally and consequently to attempt to interpret the range of our sensory experiences purely on the basis of a model that depends only on which particular afferent is firing is clearly insufficient. We cannot predict sensation simply from the nature of the stimulus. We should work backwards from the sensation through the neural apparatus to the afferent inputs and not the other way around.

Nevertheless the capacity of cutaneous afferent fibers to detect and signal particular or specific aspects of a peripheral stimulus represents a high degree of order. This order is maintained centrally by the spatial organization and morphology of the central terminals of primary afferents in the spinal cord. The dorsoventral position of an afferent terminal in the dorsal horn is determined by its physiological properties while its mediolateral and rostrocaudal position reflect the location of its peripheral terminals on the body surface. A-beta afferents terminate in Rexed laminae III to V of the dorsal horn with each type of afferent (hair follicle afferent, rapidly adapting glabrous skin afferent, Pacinian corpuscle etc.) having a characteristic morphology (Brown 1981). A-delta nociceptive afferents terminate in laminae I and V (Light and Perl 1979) while C afferents terminate partly in lamina I but mainly in lamina II (Swett and Woolf 1985, Sugiura *et al.* 1986). There is high degree of somatotopic order in the arrangement of the central terminals so that afferents that innervate contiguous skin areas in the periphery have contiguous central terminations (Swett and Woolf 1985, Woolf and Fitzgerald 1986).

Because of the greater ease with which controlled stimuli can be applied to the skin, most studies on the physiological properties of primary afferents have dealt with cutaneous afferents. Recently however, nociceptors, both thinly myelinated and unmyelinated from joints (Schaible and Schmidt 1983) and from muscle (Mense and Stahnke 1983) have been studied. Although experiments have been performed on visceral afferents (Cervero 1982) it is much more difficult to define what an "adequate" stimulus for these afferents is.

Although A-delta and C nociceptive afferents both have high thresholds and respond to noxious stimuli, there are two factors apart from conduction velocity that distinguish these afferents from each other. Firstly C fibers have the capacity to release a chemical from their peripheral terminals that directly or indirectly produce an increase in capillary permeability (neurogenic inflammation) and secondly the majority of C but not A afferents are sensitive to the neurotoxin capsaicin.

Neurogenic inflammation, also termed neurogenic edema or neurogenic extravasation has been extensively studied since its discovery over 30 years ago (Lewis 1942). Largely as a result of work by Lembeck and his colleagues (Gamse *et al.* 1980) it now appears as if the phenomenon is the result of the release by C afferents of neuropeptides such as substance P or calcitonin gene related peptide. Such release occurs whenever the C terminal is depolarized directly, via antidromic activation (the axon reflex) or indirectly via inflammatory mediators (Chahl, Solzani and Lembeck 1984). A specific function for such neurogenically mediated alteration in capillary function is not known, nor is there any evidence that the central nervous system directly controls this efferent function of C afferents. However, what the phenomenon does indicates is that in addition to detecting changes in its local environment a C afferent can also induce changes in that environment. A trophic influence of peripheral tissues on primary afferents via nerve growth factor has been shown (Levi-Montalcini, 1982). Whether afferents, via neurogenic inflammation, or other means exert a trophic influence on the periphery needs to be established. The capacity of C-primary afferents to produce neurogenic inflammation is dependent on orthograde axon transport within the afferent (Fitzgerald and Woolf 1984), while interruption of retrograde active axonal transport from the periphery induces profound chemical alterations within C afferents (Fitzgerald *et al.* 1984). This data indicates that in addition to the transfer of information along primary afferents at the millisecond time scale by the conduction of action potentials, there is also a slower conduction of information by active transport with a maximum rate only of a few mm per hour.

Capsaicin, the pungent extract of red peppers, has in common with other chemical irritants the capacity to activate chemosensitive C afferents, presumable by acting on a receptor mechanism on the afferent terminals

(Kenins 1982). However, unlike other irritants it also depolarizes C-axons (Williams and Zieglgansberger 1982) and depletes the C afferents of their neuropeptide content (see Fitzgerald 1983). When administered systemically to neonatal animals capsaicin produces a selective loss of unmyelinated axons by killing the majority of small dorsal root ganglion cells (Jancso *et al.* 1977). Capsaicin when applied directly to peripheral nerves in adult animals also produces profound changes in C fiber function (Petsche *et al.* 1983) and chemistry (Ainsworth *et al.* 1981). These results are of great interest not only because they enable us to study the reaction of the nervous system to the physical or functional absence of C fibers but also because of the possibility that capsaicin itself or a chemical derivative may provide a means of selectively reducing certain types of afferent input in man. However, as will be made clearer in a later section of this chapter, great caution will be needed because the nervous system does not respond passively to the removal of C fiber input, such a chemical deafferentation induces central compensatory changes within the dorsal horn, whose consequences in man, in terms of sensation are unknown.

Sensitization of Primary Afferents

Nociceptors are highly specific channels for the detection of noxious peripheral stimuli. By definition the threshold for activation of these sensory afferents is at a level that approaches that which damages tissue while peak firing is reached at tissue damaging intensities of peripheral stimulation. One unique property that both A-delta and C afferents share is their capacity to become sensitized. A-delta mechanoreceptive afferents for example are not normally responsive to thermal stimuli, but following repeated noxious conditioning thermal stimuli some of these afferents do become thermosensitive (Fitzgerald and Lynn 1977, Meyer and Cambell 1981). Cutaneous C-polymodal afferents also can become sensitized following a thermal noxious stimulus (Bessou and Perl 1969, Kumazawa and Perl 1977). This is not true for all C afferents so that while those that innervate hairy skin sensitize readily (La Motte *et al.* 1983) those that innervate glabrous skin do not (Meyer *et al.* 1985). Therefore mild thermal noxious stimuli have the capacity, acting on A-delta afferents in glabrous skin and C afferents in hairy skin to reduce the threshold and increase the gain of the stimulus-response relations of these afferents, directly producing or significantly contributing to hyperalgesia (defined as a reduction in the threshold and an increase in the response to noxious stimuli).

Hyperalgesia can be divided into two types, primary hyperalgesia (within the area of injury) and secondary hyperalgesia (outside the area of injury). A recent behavioral study in human volunteers has shown that thermal injuries produce both mechanical and thermal hyperalgesias (Raja *et al.* 1984). Such injuries however, while producing both primary and secondary mechanical hyperalgesia only produce a primary thermal hyperalgesia. No study on primary afferent mechanosensitive A-delta or C nociceptors has found changes in mechanical threshold or responsiveness approaching the behavioral changes observed (Meyer *et al.* 1985). The only conclusion possible is that cutaneous mechanical hyperalgesia is mediated by a mechanism other than by a change in the sensitivity of nociceptors (see later). One form of mechanical hypersensitivity that may have an important peripheral contribution however is that which is mediated by joint afferents. Following acute experimental arthritis group III (thin myelinated) and group IV (unmyelinated) joint afferents exhibit marked changes in their response properties (Schaible and Schmidt1985).

These experiments show that different afferents innervating different tissues respond in different ways to different forms of injury. While sensitization of afferents is now a clearly documented phenomenon and must contribute substantially to post-injury pain hypersensitivies, the extent, distribution and timing of the changes in the afferents indicate that such peripheral changes cannot alone account for the sensory disorders that accompany acute inflammation.

The neurochemical basis of primary afferent sensitization is of considerable interest because it provides an opportunity for pharmalogically interrupting the process. Certainly prostaglandins appear to be involved as seen by the efficacy of prostaglandin synthetase inhibitors in producing analgesia (Moncada *et al.* 1975). Other chemicals that may contribute to peripheral sensitization are bradykinin, the afferent neuropeptides such as substance P, the leucotrienes and noradrenaline (Beck and Handwerker 1974, Levine *et al.* 1984).

What still needs to be established is whether chemicals liberated by damaged tissue or by the inflammatory reaction can alter primary afferents, not by changing their transduction mechanism as with sensitization but by more subtle means. For example, if some chemical signal were taken up by peripheral terminals and transported by active transport mechanisms to the primary afferent cell body in the dorsal root ganglion where it could, via derepression of DNA, induce alterations in the metabolic activity of the neuron, this could alter the membrane properties of the afferent and its capacity to excite dorsal horn neurons. If this does occur it could represent the means by which inflammatory pain could become a neuropathic pain.

Pathological Alteration to Primary Afferent Neurones

The normal nociceptor is sensitive to the application of noxious stimuli to its receptive field. Tissue damage and acute inflammation can induce alterations in the sensitivity of such nociceptors. A major issue in terms of clinical problems is however, what pathological changes occur in the primary afferent neurone following injury to the peripheral nervous system and are these changes responsible for the sensory disorders that accompany such nerve injury.

Ten different changes can occur following damage of or injury to a primary afferent neurone:

1. death of the neurone,
2. alteration in conduction properties,
3. the production of abnormal activity,
4. pathological amplification,
5. ephaptic connections,
6. altered chemosensitivity,
7. increased mechanosensitivity,
8. altered metabolism,
9. altered trophic influence on peripheral target tissue,
10. secondary alterations in the function of the central nervous system.

The cell bodies of primary afferent neurones are located in the dorsal root ganglion and give rise to an single process which divides into a central and peripheral branch. In neonates, damage to the central and peripheral branch almost invariably leads to death of the cell body (Aldskogius *et al.* 1985) because of disruption of the active transport of growth factors. In the adult, disruption of the peripheral but not the central branch can still result in cell death (Aldskogius *et al.* 1985) but this only occurs in a minority of axotomized cells. Such cell death, in addition to depleting sensory input, could induce secondary deafferentation changes centrally.

Disturbances in conduction have long been recognized to be a fairly constant feature of injured axons, particulary when associated with demyelination (Waxman 1986). More recently it has been recognized that in pathological fibers ectopic impulses may arise from midaxon (Rasminsky 1984) or even from the cell body (Wall and Devor 1983). Of particular importance is the observation that ectopic impulses arise from sprouting neuroma fibers (Wall and Gutnick 1974) including those from C fibers (Blumberg and Janig 1984). That ectopic activity can contribute to sensory disorders such as paraesthesias have been demonstrated using microneurography (Nordin *et al.* 1984). The altered excitability of primary afferent membranes may in addition to generating spontaneous activity produce prolonged trains of impulses following normal peripheral activation of the afferent (Rasminsky 1984).

Ephaptic connections or cross talk between axons has been shown to occur in the sensory fibers of dystrophic mice (Rasminsky 1984) and between axonal sprouts in neuroma's (Seltzer and Devor 1979). The latter example may be of considerable importance in the generation of neuropathic pains if sympathetic efferents make ephaptic contacts with C afferents (Blumberg and Janig 1981).

Normal axons have little or no chemosensitivity. Following nerve injury, the collateral sprouts of a neuroma develop a sensitivity to applied or systemically administered catecholamines (Wall and Gutnick 1974, Scadding 1981). This sensitivity appears to be mediated by alpha-adrenergic receptors and may contribute to the influence of the sympathetic nervous system on certain neuropathic pains. A pathological synaptic connection between a hyperactive sympathetic efferent and an abnormally sensitive afferent could underlie certain pain states such as causalgia (Devor 1983). Whether axons develop a chemosensitivity during the inflammatory phase of some neuritides, which could cause pain, is not known.

In 1973 Wall and Gutnick demonstrated that naked sprouts of injured axons, like demyelinated axons have a highly increased mechanosensitivity. Apart from being responsible for the Tinel sign and the tenderness of a neuroma, such mechanosensitivity could contribute to the pain of nerve compression syndromes. An alternative suggestion is that the pain in such syndromes results from the activation of normal or sensitized nociceptive endings of the nervi nervori innervating inflamed or damaged nerve trunks (Asbury and Fields 1984).

Cutting the peripheral branch of an afferent neurone results, within a week, in marked alterations its metabolic activity. This includes the disappearance, both in and across the dorsal root ganglion, of marker enzymes such as fluoride resistant acid phosphatase and of the afferent neuropeptides (see Aldskogius *et al.* 1985). Although it is still not known whether this is a positive effect due to the uptake of some toxin by the damaged axon or a negative one due to interruption of the transport of essential growth factors, this finding is of considerable significance because it represents a way in which a damaged neurone can, via a change in the transmitters and neuromodulators it releases at its central end, alter dorsal horn neurones. The secondary effects of damaged afferents on the dorsal horn will be considered in a later section.

Although our knowledge is presently incomplete, injury to primary afferent neurons alters their structure, function and chemistry in a variety of different ways. Some of the reactions of the afferents may be compensatory, part of the process of repair of regeneration, others may represent a pathological disturbance resulting in abnormal function. Because neuropathic pains resulting from peripheral nerve damage are amongst the commonest cause of severe intractable clinical pain this is a field that requires intensive further investigations. What is clear though is that the damaged primary afferent neurone behaves in ways that are quite different from afferents in normal nociceptive sensory systems.

The Spinal Cord

The sensory apparatus of the spinal cord, while dependent on and modified by its afferent input, appears to process information relating to nociceptive (or physiological) pain, inflammatory pain and neuropathic pain in different, but overlapping ways.

The Dorsal Horn and Nociception

The specificity of function of primary afferents and the spatial arrangement of their central terminals provides a framework for the processing of somatosensory input into the spinal cord. Each afferent converges onto thousands of different second order neurones and each neurone receives inputs from thousands of different afferents. The receptive field properties of dorsal horn neurones are not, however, a simple reflection of the interaction between primary afferents and the dendritic trees of the neurones. If this were so then all neurones which had high threshold mechanoreceptive fields with no response to innocuous mechanical stimuli would have dendrites restricted to either laminae I and II or V with none in laminae III and IV, all neurones with C afferent inputs would have dendrites in lamina II, and the size of a neurones peripheral receptive field would be correlated with the extent of its dendritic tree. Unfortunately this is not so. The ability to record from single neurones, establish their response properties and then fill them with horseradish peroxidase (which can be demonstrated histochemically) has shown that there is no simple structure/function correlation for dorsal horn neurones (Woolf and Fitzgerald 1983, Bennett *et al.* 1984, Ritz and Greenspan 1985, Renehan *et al.* 1986, Egger *et al.* 1986). Neurones with similar morphologies have different response properties and vice versa. The missing ingredient is the simple fact that most of the inputs on dorsal horn neurones are not primary afferents but the axonal terminals of interneurones including local segmental interneurones, crossed interneurones and propriospinal interneurones (Willis 1985). Superimposed upon the afferent and interneuronal input are major descending pathways, usually but not exclusively inhibitory (Basbaum and Fields 1984).

The input to dorsal horn neurones appears to be of two different categories, fast and slow and of two different signs, excitatory and inhibitory. The fast transmitters produce brief changes in the conductance of the postsynaptic membrane following receptor binding by opening sodium or calcium ion channels or even both, and this produces a depolarization that lasts for up to 20 ms; the excitatory postsynaptic potential. The best examples of excitatory fast transmitters are the amino acids L-glutamate and L-aspartate (Jessel and Jahr 1986). However, even these transmitters have varied actions depending on which one of three receptor systems they bind to (Watkins 1986). Monosynaptic inputs appear to be mediated by quisqualate and kainate receptors while polysynaptic effects are mediated by N-methyl-D-aspartate receptors (Watkins and Evans 1981). Each receptor has different kinetics, different affinities for different agonists and antagonists and different effects on the membrane. The N-methyl-D-aspartate receptor, for example, raises intracellular calcium levels (McDermott *et al.* 1985) which may change protein kinase activity. The prototypic fast inhibitory transmitters that produce inhibitory postsynaptic potentials by hyperpolarizing the membrane are gamma amino butyric acid (GABA) and glycine. GABA also interacts with different receptors, GABA-A and GABA-B, and the interaction with the receptor is modified by the benzodiazepam receptor (Bowery 1983). Thus even the apparently simplest inputs to neurones are exceedingly complex.

A dramatic change in our appreciation of how the central nervous system operates has been the recognition that in addition to the fast acting transmitters, axon terminals release neurotransmitter/neuromodulators which have the capacity to produce changes that last for seconds, minutes or even hours. The best example of such slow transmitters in the nociceptive system is substance P. This neuropeptide is present in C-primary afferent neurones, is released centrally when C fibers are stimulated and excites dorsal horn cells (Yaksh *et al.* 1980, Di Figlia *et al.* 1982; Otsuka *et al.* 1982). However, unlike classical excitatory transmitters, substance P has a long latency before it acts and produces prolonged effects (Salt and Hill 1983).

It is extraordinarily difficult to study the biophysical changes that mediate monosynaptic afferent or polysynaptic inputs on dorsal horn neurones using intracellular recordings in vivo. A recent major breakthrough in the field has been the ability to maintain slices of the spinal cord in vitro, to record from neurones for prolonged periods and to apply drugs to the bath (Murase and Randic 1984). Using this approach Urban and Randic (1984) have shown that brief high-frequency stimulation of afferents can produce a substance P-mediated prolonged postsynaptic depolarization. This depolarization is itself subthreshold but alters the way in which the neurone responds to subsequent inputs. The significance of this finding will be discussed in the next section.

One of the most surprising aspects of dorsal horn function is that although there is a high degree of specificity of function in primary afferents the basic arrangement of dorsal horn neurones in one of convergence. While there are cells that respond exclusively to innocuous or to noxious stimuli (Willis 1985) the majority of neurones respond equally well to innocuous as well as noxious stimuli (Applebaum *et al.* 1975, Chung *et al.* 1986, Woolf and Fitzgerald 1986). These cells have been called wide dynamic range neurones or multireceptive neurones and include interneurones as well as projecting neurones.

A major problem in studying dorsal horn function is that the activity in only one neurone can be analysed at any given time (particularly when using intracellular recording techniques) and the functional context of that activity is usually unknown. Thus, if a spinothalamic cell fires in response to a noxious stimulus, it does not necessarily mean that the increased firing specifically codes the onset or magnitude of the stimulus. Such encoding is almost certainly occurring across a population of thousands of cells. Important as the particular activity in a particular neurone may be, the actual signal could be the relative activity in two neurones, or the decrease in activity of a neurone in parallel with an increase in another. Equally important are the subthreshold or subliminal inputs to neurone that are not detected by extracellular recording techniques, because relatively small changes in excitability can result in substantial changes in response profiles. The only way to fully understand the functional organization of the dorsal horn would be to be able to identify the responses, supra- and subthreshold, of all neurones to a particular input, knowing which neurones are excitatory and inhibitory interneurones, which are projecting neurones, which of the projecting axons give off collaterals that can modify local segmental activity and how the stimulus modifies descending inhibitory and excitatory influences by acting on brain stem centers. This is obviously not technically feasible. There are nevertheless two practical approaches, both necessarily indirect, that can be used to study how the dorsal horn processes sensory information. The first is that approach used by most spinal cord electrophysiologists, which is to study single neurones, to analyse as carefully as possible their receptive field properties, their morphology, the local segmental and descending inhibitory influences on the neurones, the direct and indirect actions of afferents on the cells and to correlate this information with the behavioral or sensory changes that particular stimuli elicit in the intact animal. By titrating single cell function against the behavioral responses produced by sensory inputs and by studying sufficient cells, the hope is that a pattern will emerge of how populations of neurones operate together in the spinal cord (*e.g.* Brown 1981, Willis 1985, Dubner 1985).

The alternative approach is to define the sensory processing of the spinal cord, not in terms of the activity of its components, but in terms of the biologically meaningful output it produces in response to any given input. The simplest way to do this when analysing nociceptive processing, is to analyze the flexion withdrawal reflex. In other words to look at the way noxious peripheral inputs are transformed into an output in flexor alpha motoneurones that results in the contraction of flexor muscles that will remove a limb away from the stimulus. The advantages of this approach are, firstly, that it is relatively easy to measure activity in flexor motoneurones without disturbing the spinal cord, and secondly, that in man there is a close correlation between the perception of noxious stimuli and the activation of flexor motoneurones (Willer *et al.* 1979).

Flexor alpha motoneurones in the decerebrate-spinal rat preparation are characterized by an absent spontaneous discharge, high threshold polymodal cutaneous receptive fields and inputs from A and C afferent fibers (Woolf and Swett 1984, Cook and Woolf 1985). The threshold for eliciting activation of the motoneurones is in the noxious range and the response adapts rapidly to sustained peripheral stimuli. These properties represent the electrophysiological correlate of the high threshold phasic flexion withdrawal reflex observed behaviorally. Repeated standard cutaneous stimuli, provided they do not injure the skin, evoke a remarkably stable response measured in terms of the number of action potentials elicited in the motoneurones (Woolf 1983). These results show, that in spite of the convergence of innocuous and noxious inputs seen on most dorsal horn neurones, a nociceptive afferent input is transformed by the spinal cord into a specific output that cannot be elicited by low threshold inputs. In addition, the stability of flexion reflex provides a particulary useful way of measuring any disturbance or alteration in the functional performance of the spinal cord.

Acute Inflammation and the Spinal Cord

The chronic decerebrate rat provides a very convenient model for studying the effects of tissue damaging stimuli on the function of the spinal cord. These animals have intact brain stem and spinal reflexes and respond to noxious stimuli in a way that is qualtitatively identical to that found in intact animals; a brisk rapid flexion withdrawal, orientation of the body to the site of the stimulus, a coordinated escape response and vocalization (Woolf 1984). Acute injury, either thermal or chemical, induces a major change in the reflex performance of the animal. The threshold falls to innocuous levels and the responsiveness is greatly exaggerated (Woolf 1984). The injury transforms the reflex from a high threshold phasic response to a low threshold tonic one. The resemblance of these changes to the postinjury pain hypersensitivities in man, where threshold and responsiveness also change, make this an interesting model for studing of the effects of peripheral inflammatory lesions on the function of the nervous systems.

Electrophysiological recordings from a flexor motoneurone before and after an acute injury to its receptive field show in addition to the fall in threshold, an expansion of the size of the receptive field and a greatly increased response to standard stimuli (Woolf 1983). Once the injury induced changes are established, injecting local anaesthetic into the site of the inflamed tisssue to produce a sensory block does not result in an immediate return on the flexor motoneurones response properties to the preinjury level (Woolf 1983). This indicates that some afferent signal generated by the injury triggers an alteration in the functional performance of the spinal cord. When the properties of individual primary afferents, before and after an injury, are compared with the changes in the flexor motoneurones no changes in the input signal can be found that can account for the change in the output signal (Woolf and McMahon 1985).

More direct evidence for the capacity of afferent input to generate prolonged changes centrally, within the spinal cord, have come from experiments that demonstrate that brief (up to 20 s) conditioning stimuli at low frequencies (1 Hz) can produce prolonged (up to 90 minutes) increases in the excitability of the flexion reflex (Wall and Woolf 1984, Woolf and Wall 1986). Such increases in afferent-induced excitability only occur if the stimulus strength is sufficient to activate unmyelinated afferents. Therefore C fiber input to the spinal cord, in addition to producing an input concerning the onset, location and duration of peripheral noxious stimuli, also produces a prolonged increase in the excitability of the spinal cord. That this phenomenon may be mediated by neuropeptides is seen by two experiments. Firstly, when a peripheral nerve is axotomized, within one week the neuropeptide content disappears but the capacity of the C afferents to excite the spinal cord remains intact. However these axotomized C afferents are unable to produce prolonged facilitation of the flexion reflex (Wall and Woolf 1986). Secondly the intrathecal injection of low doses of substance P and calcitonin gene-related peptide produce prolonged increases in the excitability of the spinal cord (Woolf and Wiesenfeld-Hallin 1986). Of some interest, in view of the clinical differences between superficial and deep injuries, is the finding that C fibers innervating different target tissues have different efficacies in producing central facilitation. Cutaneous C afferents, for example, are much less effective than those that innervate joints or muscle (Woolf and Wall 1986).

The prolonged facilitatory effects of C afferent inputs are not the result of changes in the membrane properties of the flexor motoneurones or of the afferent central terminals (Cook *et al.* 1986). The changes must be occurring somewhere in the chain of interneurones that link the afferent input with the motor output. Recently C afferent conditioning stimuli have been shown to produce profound changes in the receptive field properties of dorsal horn neurones including those that send axons to higher centres (Cook *et al.* 1987). These changes include the expansion of receptive field size, increases in spontaneous activity and increases in response to standard stimuli. After the conditioning input some neurones that had originally responded only to noxious stimuli began to respond to noxious and to innocuous stimuli.

Although these results must be interpreted with caution, they may have a bearing on the pathogenesis on inflammatory pain hypersenstivities. The pain and tenderness that occours in man after tissue injury may result not only from changes in the transduction properties of nociceptors (peripheral sensitization) but also from a central sensitization produced by a C afferent barrage. The excitability increase produced by such a C-input may lead to A-beta afferent inputs being able to activate the flexion reflex and to produce pain. This dynamic functional plasticity transfers a sherringtonian "labeled line" nociceptive system into one where the lines become functionally crossed. If the animal data is indicative of changes that occur in man then the processing of information that mediates physiological pain is qualitatively different from that which mediate inflammatory pain. One set of experiments that is compatible with this, is the finding that low doses of morphine have the capacity to completely block the prolonged C fiber-induced excitations without substantially modifying the response to noxious stimuli (Woolf and Wall 1986). This pharmacological dissociation of nociception and injury or C fiber-induced central changes indicates that attempts to manage inflammatory pain should be aimed centrally as well as peripherally.

Neuropathic Pain and the Spinal Cord

In the section of this chapter devoted to pathological changes in primary afferent neurones following injury, three different kinds of alterations were discussed; 1. those that result in an abnormal afferent input, 2. those that are associated with a change in the metabolic activity and chemical composition of the afferent and 3. those that result in cell death. Assuming that peripheral nerve injury produces no secondary changes within the central nervous system, the first and last category of changes would, nevertheless, result in abnormal sensory processing by virtue of positive changes in the first example and negative in the last. A convincing case has been made that some of the characteristic neuropathic pains that result from nerve injury arise solely from abnormalities in input due to ectopic activity, ephaptic connections and prolonged discharges (Ochoa *et al.* 1985). There is, however, convincing evidence that peripheral nerve injury does induce alterations in the central nervous system (Wall 1983) and consequently any attempt to understand the pathogenesis of neuropathic pain must take this into account.

Injury to the peripheral branch of a primary afferent neurone results in transganglionic changes in the central branch including the axonal terminals within the dorsal horn. Ultrastructural evidence points to at least some transganglionic degeneration (Aldskogius *et al.* 1985) but whether this is due to cell death or to removal of a trophic influence remains controversial. There is no morphological evidence for collateral sprouting in the dorsal horn following nerve section but transynaptic changes in the morphology of some neurones has been found (Gobel 1984). Electrophysiological evidence has shown that peripheral nerve injury results in a decrease both in pre- and postsynaptic inhibitory mechanisms (Wall and Devor 1981, Woolf and Wall 1982). The functionally denervated area of the spinal cord begins, after several days, to show alterations such as an expansion in dorsal horn neuronal receptive field size so that cells that were formerly innervated exclusively by the denervated skin areas start to respond to distant areas (Devor and Wall 1981). Although the mechanism of these changes is not known (one suggestion is the activation of normally suppressed or ineffective synapses) what these results do show is that interrupting afferent input does not simply result in functionally silent areas within the central nervous system. Compensatory mechanism change the performance of the central nervous system. This central plasticity may under normal circumstances represent a means to deal with minor injuries and disturbances but cannot adequately compensate for major disturbances and instead introduces a new, secondary pathology, to the system. While this is, in its own right, of interest as an example of plasticity in the nervous system, the practical implications for neurosurgical intervention to the somatosensory system are immense. Interrupting a sensory pathway may alleviate a sensory disorder by blocking an abnormal afferent signal but it may also activate changes proximal to the site of the surgery, generating a new disorder.

One well documented example of deafferentation pain is that of dorsal root avulsion as in brachial plexus avulsions (Wynn Parry 1980). Experiments in animal models have shown that dorsal root lesions result in a self-mutilation behavior aimed at the boundary between the anesthetic area and the normal skin (Wall 1983). This behavior is associated with spontaneous, irregular and abnormal hyperactivity within the deafferented spinal cord (Lombard and Larabi 1983, Wall *et al.* 1979, Mendell *et al.* 1977). The receptive fields of dorsal horn neurons, when present, are abnormal and expanded (Basbaum and Wall 1976). With time, similar changes also appear in the thalamus (Lombard and Larabi 1983).

Lesions directed to the central nervous system, such as anterolateral cordotomy, can produce self-mutilation behavior similar to that found as part of the peripheral nerve deafferentation syndrome (Levitt and Levitt 1981), implying that these central lesions produce disturbances resembling those found after dorsal root or periperal nerve section. Certainly stimulation of the anterolateral columns of patients with chronic neuropathic pain produce sensations that are never found in those with nociceptive or inflammatory pain (Tasker *et al.* 1983).

From a general point of view neuropathic pain may be thought of arising in one of two ways. The pain may result from abnormal compensatory mechanisms in response to deafferentation producing an imbalance in the functional circuitry of the somatosensory system; alternatively, a constant barrage of abnormal inputs may induce permanent changes in the circuitry of the nervous system. There is evidence for the first mechanism. Experiments need to be performed to see if the second is also true.

One indication that neuropathic pains are not just negative phenomena due to deafferentation, is the frequent observation that such pains are often accompanied by abnormalities in the sympathetic nervous system. Suppression of sympathetic innervation by surgical sympathectomy (Spurling 1930) local sympathetic blocks (Livingstone 1938) and by regional pharmacological blocks (Hannington-Kiff 1983) are amongst the most effective means available for treating such disorders. There is no evidence, however, that hyperactivity of the sympathetic nervous system is itself responsible for producing pain (see Ochoa *et al.* 1985) instead it appears as if a sympathetic-afferent interaction in the periphery or at the side of the nerve injury, acting either by ephapsis or by abnormal synaptic connections may be responsible (Devor 1983). Abnormal peripheral connections, cannot however be the sole explanation. What, for example, causes the abnormal sympathetic outflow (Blumberg and Janig 1983)? It must be an abnormal afferent input acting on the spinal cord changing the response properties of dorsal horn neurones and preganglionic sympathetic motoneurones. This has the consequence, as Livingstone realized in 1948, of generating a vicious circle with the sympathetic efferent outflow modifying the afferent input that changes the sympathetic outflow. There is no reason to believe that these abnormal transformations do not also alter the functional performance of those neurones that transmit sensory informations to the thalamus. These changes in the spinal cord appear in some cases to be irreversible because surgery, such as excision of a neuroma for causalgia, does not necessarily alleviate the disorder (Noordenbos and Wall 1981).

Conclusions

This Chapter has reviewed how the primary afferent neurone and the neurones in the spinal cord may contribute to the pathogenesis of nociceptive, inflammatory and neuropathic pain. While our knowledge remains incomplete, experimental work continues to contribute to a greater understanding of the normal and disturbed somatosensory system and this must inevitably lead to more effective measures to manage the different kinds of pain.

The somatosensory system does not end in the spinal cord but, while we have some information on nociceptive processing in the thalamus and cortex (Willis 1985), practically nothing is known about their roles in pathological pain (Benoist *et al.* 1985). Another important component of the somatosensory system, complementing the ascending pathways, is descending inhibitory traffic from the brain stem nuclei such as the midline raphe nuclei and the ventrolateral reticular formation (Basbaum and Fields

1984). Although their capacity to modulate sensory transmission through the spinal cord by pre- and postsynaptic mechanism is well documented, as are the neurotransmitter systems involved (serotonin, noradrenaline and the endogenous opioids), what is quite unknown is how the descending inhibitory pathways are activated or deactivated during clinical conditions. Utilization of endogenous inhibitory systems by electrical stimulation (*e.g.*, Transcutaneous Electrical Nerve Stimulation for afferent inhibition or brain stimulation to activate the descending pathways) or by pharmacological means (*eg.*, morphine to mimic the opioid peptides) are topics that require separate chapters.

The important messages to emerge from recent research into the pathogenesis of pain are that the processes responsible for the different types of pain are not equivalent, that inflammatory and neuropathic pains are likely to have central as well as peripheral causes, and that because of the central component and the capacity of the nervous system for plastic changes, surgical interruption of sensory pathways may not always be the most appropriate form of intervention.

References

1. Ainsworth A, Hall P, Wall PD, Allt G, MacKensie ML, Gibson S, Polak JM (1981) Effects of capsaicin applied locally to adult peripheral nerve. II. Anatomy and enzyme and peptide chemistry of peripheral nerve and spinal cord. Pain 11: 379–388
2. Aldskogius H, Arvidsson J, Grant G (1985) The reaction of primary sensory neurones to peripheral nerve injury with particular emphasis on transganglionic changes. Brain Res Reviews 10: 27–40
3. Applebaum AE, Beall JE, Foreman RD, Willis WD (1975) Organization and receptive fields of primate spinothalamic tract neurons. J Neurophysiol 38: 572–586
4. Asbury HK, Fields HL (1984) Pain due to peripheral nerve damage: A hypothesis. Neurology (Minneap) 34: 1587–1590
5. Basbaum AI, Fields HL (1984) Endogenous pain control systems: Brainstem spinal pathways and endorphin circuitry. Ann Rev Neurosci, 7: 309–338
6. Basbaum AI, Wall PD (1976) Chronic changes in the response of cells in cat dorsal horn, following partial deafferentation. The appearance of responding cells in a previous non-responding region. Brain Res 116: 181–204
7. Beck PW, Handwerker HO (1974) Bradykinin and serotonin effects on various types of cutaneous nerve fibres. Pflügers Arch 347: 209–222
8. Bennett GJ, Nishikawa N, Lu G-U, Hoffert MJ, Dubner R (1984) The morphologie of dorsal column postsynaptic spinomedullary neurons in the cat. J Comp Neurol 224: 568–578
9. Benoist JM, Kayser V, Gaitron M, Guilbaud G (1985) Changes in responses of ventrobasal thalamic neurons during carrageenin-induced inflammation in the rat. Adv in Pain Res and Ther 9: 295–304

10. Bessou P, Perl ER (1969) Response of cutaneous sensory units with unmyelinated fibers to noxious stimuli. J Neurophysiol 32: 1025–1043
11. Blumberg H, Janig W (1981) Neurophysiological analysis of efferent sympathetic and afferent fibers in skin nerves with experimental produced neuromata. In: Siegfried J, Zimmermann M (eds) Phantom and stump pain. Springer, Berlin Heidelberg New York, pp 15–31
12. Blumberg H, Janig W (1983) Changes of reflexes in vasoconstrictor neurons supplying the cat hindlimb following chronic nerve lesions: a model for studying mechanisms of reflex sympathetic dystrophy? J Auton Nerv Syst 7: 399–411
13. Blumberg H, Janig W (1984) Discharge patterns of afferent fibres from a neuroma. Pain 20 (4): 335–353
14. Brown AG (1981) Organization in the spinal cord; the anatomy and physiology of identified neurones. Springer, Berlin Heidelberg New York
15. Bowery NG (1983) Classification of GABA receptors. In: Erna SJ (ed) The GABA receptors. Humana Press, New Jersey, pp 178–213
16. Burgess PR, Perl ER (1973) Cutaneous mechanoreceptors and nociceptors. In: Iggo A (ed) Handbook of sensory physiology, somatosensory system, vol 2. Springer, Berlin Heidelberg New York, pp 29–78
17. Campbell JN, la Motte RH (1983) Latency to detection of first pain. Brain Res 266: 203–208
18. Cervero F (1982) Afferent activity evoked by natural stimulation of the biliary system in the ferret. Pain 13: 137–151
19. Chahl LA, Szolcsanyi J, Lembeck F (1984) Neurogenic inflammation. Akademiai Kiado, Budapest
20. Chung JM, Surmeier DJ, Lee KH, Sorkin LS, Honda CN, Tsang Y, Willis WD (1986) Classification of primate spinothalamic and somatosensory thalamic neurons based on cluster analysis. J Neurophysiol 56: 308–327
21. Cook AJ, Woolf CJ (1985) Cutaneous receptive field and morphological properties of hamstering flexor alphamotorneurone in the rat. J Physiol 364: 249–263
22. Cook AJ, Woolf CJ, Wall PD (1981) Prolonged C-fibre mediated facilitation of the flexion reflex is not due to changes in afferent terminal or motoneurone excitability. Neurosci Lett 70: 91–96
23. Cook AJ, Woolf CJ, Wall PD, McMahon SB (1987) Dynamic receptive field plasticity in rat spinal dorsal horn following C-primary afferent input. Nature 325: 151–153
24. Devor M (1983) Nerve pathophysiology and mechanism of pain in causalgia. J Auton Nerv Syst 7: 371–384
25. Devor M, Wall PD (1981) Plasticity in the spinal cord sensory map following peripheral nerve injury in rats. J Neurosci 1: 679–684
26. Di Figlia M, Aronin N, Leeman SE (1982) Light microscopic and ultrastructural localization of immunoreactive substance P in the dorsal horn of monkey spinal cord. Neuroscience 7: 1127–1139
27. Dubner R (1985) Specialization in nociceptive pathways: sensory discrimination, sensory modulation and neural connectivity. Adv Pain Res and Ther 9: 11–138

28. Egger MD, Feeman NCG, Jacquin M, Proshanski E, Semba K (1986) Dorsal horn cells in the cat responding to stimulation of the plantar cushion. Brain Res 383: 68–83
29. Fitzgerald M (1983) Capsaicin and sensory neurones—a review. Pain 15: 109–130
30. Fitzgerald M, Lynn B (1977) The sensitization of high threshold mechanoreceptors with myelinated axons by repeated heating. J Physiol 265: 549–563
31. Fitzgerald M, Woolf CJ (1984) Axon transport and sensory C-fibre function. In: Chahl CA, Szolcsanyi J, Lembeck F (eds) Neurogenic inflammation. Akademiai Kiado, Budapest, pp 119–140
32. Fitzgerald M, Woolf CJ, Gibson SJ, Mallaburn PS (1984) Alterations in the structure, function and chemistry of C fibers following local application of vinblastine to the sciatic nerve of the rat. J Neurosci 4: 430–441
33. Gamse R, Holzer P, Lembeck F (1980) Decrease of substance P in primary afferent neurones and impairment of neurogenic plasma extravasation by capsaicin. Br J Pharmacol 68: 207–213
34. Georgopoulos AP (1977) Stimulus-response relations in high-threshold mechanothermal fibers innervating primate glabrous skin. Brain Res 128: 547–552
35. Gobel S (1984) Trans-synaptic effects of peripheral nerve injury. J Neurosci 4: 2281–2290
36. Gybels J, Handwerker HO, van Hees J (1979) A comparison between the discharges of human nociceptive fibers and the subject's ratings of his sensations. J Physiol 292: 193–206
37. Hannington-Kiff, JG (1983) Hyperadrenergic-effected limb causalgia; Relief by pharmacologic norepinephrine blockade. Am Heart J 103: 152–159
38. Iggo A, Ogawa H (1971) Primate cutaneous thermal nociceptors. J Physiol 216: 77P–78P
39. Jancso G, Kiraly E, Jancso-Gabor A (1977) Pharmacologically induced selective degeneration of chemosensitive primary sensory neurones. Nature Lond 270: 741–743
40. Jessell TM, Jahr CE (1986) Synaptic interactions between dorsal root ganglia and dorsal horn neurones in cell culture: amino acids, nucleotides and peptides as possible fast and slow excitatory transmitters. In: Iversen LL, Goodman E (eds) Fast and slow chemical signalling in the nervous system. Oxford University Press, Oxford
41. Kenins P (1982) Responses of single nerve fibers to capsaicin applied to the skin. Neurosci Lett 29: 311–316
42. Konietzny F, Perl ER, Trevino D, Light A, Hensel H (1981) Sensory experiences in man evoked by intraneural electrical stimulation of intact cutaneous afferent fibers. Exp Brain Res 42: 219–222
43. Kumazawa T, Pearl ER (1977) Primate cutaneous sensory units with unmyelinated (C) afferent fibers. J Neurophysiol 40: 1325–1338
44. La Motte RH, Thalhammer JG, Robinson CJ (1983) Peripheral neural correlates of magnitude of cutaneous pain and hyperalgesia: a comparison of neural events in monkey with sensory judgements in human. Neurophysiol 50: 1–26

45. Levi-Montalcini R (1982) Developmental neurobiology and the natural history of nerve growth factor. Ann Rev Neurosci 5: 341–362
46. Levine JD, Lau W, Kwiat G, Goetzl EJ (1984) Leukotriene B4 produces hyperalgesia that is dependent on polymorphonuclear leukocytes. Science 225: 743–745
47. Levitt M, Levitt JH (1981) The deafferentation syndrome in monkeys: dysesthesias of spinal origin. Pain 10: 129–147
48. Lewis T (1942) Pain. Macmillan, New York
49. Light AR, Perl ER (1979) Spinal termination of functionally identified primary afferent neurons with slowly conducting myelinated fibers. J Comp Neurol 186: 133–150
50. Livingstone K (1938) Phantom limb pain; report of 10 cases in which it was treated by injections of procaine hydrochloride near thoracic sympathetic ganglions. Arch Surg 37: 353–370
51. Livingstone K (1948) The vicious circle in causalgia. Ann NY Acad Sci 50: 247–258
52. Lombard MC, Larabi Y (1983) Electrophysiological study of cervical dorsal horn cells in partially deafferented rats. In: Bonica JJ, Lindblom U, Iggo A (eds) Advances in pain res and ther, vol 5. Raven Press, New York, pp 147–154
53. MacDermott AB, Mayer ML, Westbrook GL, Smith DJ, Barber JL (1986) NMDA-receptor activation increases cytoplasmic calcium concentrations in cultured spinal cord neurones. Nature 321: 519–521
54. Mendell LM, Sassoon EM, Wall PD (1977) Properties of synaptic linkage from distant afferents onto dorsal horn neurons in normal and chronically deafferented cats. J Physiol (Lond) 285: 299–310
55. Mense DER, Stahnke M (1983) Responses in muscle afferent fibres of slow conduction velocity to concentrations and ischaemia in the cat. J Physiol 342: 383–397
56. Meyer RA, Cambell JN (1981) Myelinated nociceptive afferents account for the hyperalgesia that follows a burn to the hand. Science 213: 1527–1529
57. Meyer RA, Cambell JN, Raja SN (1985) Peripheral neural mechanism of cutaneous hyperalgesia. Adv Pain Res and Ther 9: 53–72
58. Moncada S, Ferreira SH, Vane JR (1975) Inhibition of prostaglandin biosynthesis as the mechanism of analgesia of aspirin-like drugs in the dog knee joint. Eur J Pharmacol 31: 250–260
59. Murase K, Randic M (1983) Electrophysiological properties of rat spinal dorsal horn neurons in vitro: calcium dependent action potenitals. J Physiol (Lond) 334: 141–153
60. Noordenbos W, Wall PD (1981) Implications of the failure of nerve resection and graft to cure chronic pain produced by nerve lesions. J Neurol Neurosurg Psychiat 44: 1068–1073
61. Nordin M, Nystrom B, Wallin U, Hagbarth KE (1984) Ectopic sensory discharges and paraesthesiae in patients with disorders of peripheral nerves, dorsal roots and dorsal columns. Pain 20: 231–245

62. Ochoa JL, Torebjork E, Marchetti P, Swok M (1985) Mechanisms of neuropathic pain: cumulative observations, new experiments and further speculation. Adv Pain Res Ther 9: 431–450
63. Otsuka M, Konishi S, Yanagisawa M, Tsunoo A, Akagi H (1982) Role of substance P as a sensory transmitter in spinal cord and sympathetic ganglia. CIBA Foundation Symp 91: 13–34
64. Perl ER (1985) Unravelling the story of pain. Adv Pain Res and Ther 9: 1–30
65. Petsche U, Fleischer E, Lembeck F, Handwerker HO (1977) The effect of capsaicin application to a peripheral nerve on impulse conduction in functionally identified afferent nerve fibres. Brain Res 122: 534–540
66. Raja S, Cambell JN, Meyer RA (1984) Evidence for different mechanism of primary and secondary hyperalgesia following heat injury to the glabrous skin. Brain 107: 1179–1188
67. Rasminsky M (1984) Ectopic impulse generation in pathological nerve fibres. In: Dyck PJ, Thomas PK, Lambert EH (eds) Peripheral neuropathology, 2nd edn. WB Saunders, Philadelphia
68. Renehan WE, Jacquin MF, Mooney RD, Rhoades RW (1986) Structure-function relationships in rat medullary and cervical dorsal horns. II. Medullary dorsal horn cells. J Neurophysiol 55: 1187–1201
69. Ritz LA, Greenspan JD (1985) Morphological features of lamina V neurons receiving nociceptive input in cat sacrocaudal spinal cord. J Comp Neurol 238: 440–452
70. Salt TE, Hill RG (1983) Neurotransmitter candidates of somatosensory primary afferent fibres. Neuroscience 10: 1083–1103
71. Scadding JW (1981) Development of ongoing activity, mechanosensitivity and adrenaline sensitivity in severed peripheral nerve axons. Exp Neurol 73: 345–364
72. Schaible HG, Schmidt RF (1983) Activatin of groups III and IV sensory units in medial articular nerve by local mechanical stimulation of knee joint. J Neurophysiol 49: 35–44
73. Schaible HG, Schmidt RF (1985) Effects of an experimental arthritis on the sensory properties of fine articular afferent units. J Neurophysiol 54: 1109–1121
74. Seltzer M, Devor M (1979) Ephaptic transmission in chronically damaged peripheral nerves. Neurology (Minneap) 29: 1061–1064
75. Sherrington CS (1981) The integrative action of the nervous system. Yale University Press, New Haven 1906, 2nd ed 1947, Yale paperbound
76. Spurling RG (1930) Causalgia of upper extremity, treatment by dorsal sympathetic ganglionectomy. Arch Neur Psychiat 23: 784–788
77. Sugiura Y, Lee CL, Perl ER (1986) Central projections of identified, unmyelinated (C) afferent fibres innervating mammalian skin. Science 234: 336–358
78. Swett J, Woolf CJ (1985) The somatotopic organization of primary afferent terminals in the superficial laminae of the dorsal horn of the rat spinal cord. J Comp Neurol 231: 66–77
79. Tasker RR (1984) Deafferentation. In: Wall PD (ed) Textbook of pain. Churchill Livingstone, Edinburgh, pp 119–132

80. Tasker RR, Tsuda T, Hawrylysyn P (1983) Clinical neurophysiological investigation of deafferentation pain. Adv Pain Res and Ther 5: 713–738
81. Torebjork HE, Ochoa JL (1983) Selective stimulation of sensory units in man. Adv Pain Res and Ther 5: 99–104
82. Torebjork HE, la Motte RH, Robinson CJ (1984) Peripheral neural correlates of magnitude of cutaneous pain and hyperalgesia: Simultaneous recordings in humans of sensory judgements of pain and evoked responses in nociceptors with C-fibres. J Neurophysiol 51: 325–339
83. Urban L, Randic M (1984) Slow excitatory transmission in rat dorsal horn: possible mediation by peptides. Brain Res 290: 336–341
84. Wall PD (1983) Alterations in the central nervous system after deafferentation: Connectivity control. Adv Pain Res and Ther 5: 677–689
85. Wall PD, Devor M (1981) The effects of peripheral nerve injury on dorsal root potentials and transmission of afferent signals into the spinal cord. Brain Res 209: 95–111
86. Wall PD, Devor M (1983) Sensory afferent impulses originate from dorsal root ganglia as well as from the periphery in normal and nerve injured rats. Pain 17: 321–339
87. Wall PD, Gutnick M (1974) Ongoing activity in peripheral nerves: the physiology and pharmacology of impulses originating from a neuroma. Exptl Neurol 43: 580–593
88. Wall PD, McMahon SB (1986) The relationship of perceived pain to afferent nerve impulses. TINS
89. Wall PD, Devor M, Inbal R *et al* (1979) Autonomy following peripheral nerve lesions: experimental anaesthesia dolorosa. Pain 7: 103–111
90. Wall PD, Woolf CJ (1984) Muscle but not cutaneous C-afferent input produced prolonged increases in the excitability of the flexion reflex in the rat. J Physiol 356: 443–458
91. Wall PD, Woolf CJ (1986) The brief and the prolonged facilitatory effects of unmyelinated afferent input on the rat spinal cord are independently influenced by peripheral nerve section. Neuroscience 17: 1199–1205
92. Watkins JC (1986) Selective antagonists define sub-classes of excitatory amino acid receptors. In: Iversen LL, Goodman E (eds) Fast and slow chemical signalling in the nervous system. Oxford University Press, Oxford
93. Watkins JC, Evans RH (1981) Excitatory amino acid transmitters. Ann Rev Pharmacol Toxicol 21: 165–204
94. Waxman SG (1986) Normal and abnormal axonal projections. In: Asbury AK, McKhann GM, McDonald WI (eds) Nervous system diseases. Heinemann, London, pp 36–57
95. Williams JT, Zieglgansberger W (1982) The acute effects of capsaicin on rat primary afferents and spinal neurons. Brain Res 253: 125–131
96. Willer JC, Boureau F, Albe-Fessard D (1979) Supraspinal influences on nociceptive flexion reflex and pain sensation in man. Brain Res 179: 61–68
97. Willis WD (1985) The pain system. The neural basis of nociceptive transmission in the mammalian central nervous system. S Karger, Basel
98. Woolf CJ (1983) Evidence for a central component of postinjury pain hypersensitivity. Nature 308: 686–688

99. Woolf CJ (1984) Long term alterations in the excitability of the flexion reflex produced by peripheral tissue injury in the chronic decerebrate rat. Pain 18: 325–343
100. Woolf CJ, Fitzgerald M (1983) The properties of neurones recorded in the superficial dorsal horn of the rat spinal cord. J Comp Neurol 221: 313–328
101. Woolf CJ, Fitzgerald M (1986) The somatotopic organization of cutaneous afferent terminals and dorsal horn neuronal receptive fields in the superficial and deep laminae of the rat lumbar spinal cord. J Comp Neurol 251: 517–531
102. Woolf CJ, McMahon SB (1985) Injury-induced plasticity of the flexor reflex in chronic decerebrate rats. Neuroscience 16: 395–404
103. Woolf CJ, Swett JE (1984) The cutaneous contribution to the hamstring flexor reflex in the rat: an electrophysiological and anatomical study. Brain Res 303: 299–312
104. Woolf CJ, Wall PD (1982) Chronic peripheral nerve selection diminishes the primary afferent A-fibre mediated inhibition of rat dorsal horn neurones. Brain Res 242: 77–85
105. Woolf CJ, Wall PD (1986) Morphine sensitive and morphine insensitive actions of C-fibre input on the rat spinal cord. Neuroscience Lett 64: 221–225
106. Woolf CJ, Wall PD (1986) The relative effectiveness of C primary afferent fibres of different origins in evoking a prolonged facilitation on the flexor reflex in the rat. J Neurosci 6: 1433–1442
107. Woolf CJ, Wiesenfeld-Hallin Z (1986) Substance P and calcitonin gene related peptide synergistically modulate the gain of the nociceptive flexor reflex in the rat. Neurosci Lett 66: 226–230
108. Wynn Parry CB (1980) Pain in avulsion lesions of the brachial plexus. Pain 9: 41–53
109. Yaksh TL, Jessell TM, Gamse R, Mudge AW, Leeman SE (1980) Intrathecal morphine inhibits substance P release from mammalian spinal cord in vivo. Nature 286: 155–157

Spinal Cord Stimulation for Spasticity

J. GYBELS[1] and D. VAN ROOST[2]

[1] Kliniek voor Neurologie en Neurochirurgie, Universitair Ziekenhuis "Gasthuisberg", Leuven (Belgium)
[2] Neurochirurgische Universitätsklinik, Bonn-Venusberg (Federal Republic of Germany)

With 6 Figures

Contents

I. Spinal Cord Stimulation

1. Historical Review

a) Pain as the First Application of Spinal Cord Stimulation

Spinal cord stimulation (SCS) was introduced in 1967 by Shealy, Mortimer and Reswick as a means for obtaining relief from chronic pain.

At that time, this approach seemed to be fully supported by the gate-control theory of pain, formulated by Melzack and Wall in 1965. According to an updated version of this theory (Melzack and Wall 1970) impulses conducted over large afferent fibers and over fibers from higher levels activate cells of the substantia gelatinosa, which in turn affect the—nociceptive—small fiber input by presynaptic inhibition.

The prerequisite for a successful pain relief by SCS turned out to be an electric current application directly over the involved spinal segment, which in the classical case of electrode placement over the dorsal surface of the spinal cord coincides with the generation of paresthesias in the painful area. This circumstance and its anatomical counterpart gave rise to the term "dorsal column stimulation (DCS)".

Pain reduction without paresthesias however, for instance by placement of the electrodes over the anterior cord surface, has been reported as well (Larson *et al.* 1974, Campbell 1981), and therefore has challenged the mediating role of the dorsal columns or even of the lemniscal pathways generally in the neurophysiological basis of spinal stimulation for pain.

The prolonged relief from pain following cessation of current application is another intriguing feature of SCS, that could not be satisfactorily explained in terms of the gate-control theory.

The remaining uncertainty about the real target of the current applied for pain alleviation within the cord, led to the use of "spinal cord stimulation (SCS)" as a generally accepted designation of the technique instead of the earlier term "dorsal column stimulation".

SCS was adopted with enthusiasm by many neurosurgeons, and the more recent, percutaneously implantable, form was also adopted by non-neurosurgical pain clinics. The method offered an elegant new alternative to patients with intractable pain, that from a technical point of view was relatively simple, at least compared with stereotactic procedures, and from a physiological, ethical and forensic point of view, nondestructive and hence testable on a reversible basis.

The target group included patients suffering from chronic low back pain after so-called failed back surgery, phantom pain, postherpetic neuralgia, cramping pain in multiple sclerosis, pain due to peripheral vasculopathic ischemia, cancer pain, ect. Good to excellent long-term results were reported by Siegfried and Hood in 45 to 73% of the cases (Siegfried and Hood 1983, Long *et al.* 1981).

b) Other Applications

The application of SCS for pain as a problem common to a large variety of disease led to interesting marginal observations, such as the healing of arteriosclerotic cutaneous ulcers and a significant increase in peripheral blood flow in nonadvanced vasculopathy. A decrease in sympathetic activity was postulated as the responsible working mechanism.

Cook and Weinstein reported in 1973 on the fortuitous application of SCS for low back ache in a paraplegic women who was affected by multiple sclerosis for some years. After a delay of some days she was able again to raise her legs off the bed, which had been impossible before. This surprising favorable effect on motor impairment was reproducible after SCS withdrawal in the same patient and in one out of two other, similar cases.

In 1976 Cook reported good results in a large majority (90%) of 70 other MS patients treated with SCS since those first trials. He recommended the method mainly for selected MS cases with proven diagnosis and moderate impairment, left, however, the indication open to other dystonic or dyskinetic conditions, and mentioned, moreover, improvements in peripheral blood circulation, respiration and micturition. He observed SCS effects that extended far beyond the immediate area of application, such as an improved arm function by stimulation at the midthoracic level.

2. *Technical Development—Stimulation Parameters*

The implantation technique of SCS does not substantially differ whether pain or dystonia is the indication for the procedure.

In the early years, a laminectomy in the thoracic area was needed in order to place bipolar electrodes extra-, endo- or intradurally over the midline of the spinal cord. Connections werde made with a device implanted subcutaneously, usually in the subcostal or subclavicular region. The first, and most widely used, radiofrequency-linked arrangement allowed the electric power supply to be transmitted to the electrodes externally, by placing the coil antenna of a transmitter device over the skin covering the implanted receiver. The external miniaturized battery-powered transmitter generates the rectangular electric pulses and permits variation in different parameters within certain ranges.

The reasons for uncoupling electrodes from supply and control are to be found in the high power requirements of SCS, about 2 orders of magnitude greater than those for cardiac pacing, and moreover in a demand for maximal adjustability of stimulation parameters.

Cathodal stimulation emerged as the most effective and energy-saving method (Ranck 1975, Sherwood 1981, Davis and Gray 1981, Law 1983). As for polarity, the claim of free choice could already be omitted. The remaining variables were voltage, frequency and pulse width, respectively, in ranges for pain therapy, from 1–10 volt, 20–200 Hz and 100–500 μs. For the specific purpose of controlling ataxia, however, Cook mentioned in 1976 the usefulness of higher frequency stimulation, without further specification.

Since 1973, the less invasive percutaneous installation of wire leads into the epidural space by means of a puncture needle (type "Tuohy-Huber" or "Hustead") found wide acceptance especially as a preliminary selection procedure. It can be performed under local anesthesia and fluoroscopic control. The percutaneously implanted wire electrode is not primarily attached to any intraspinal structure and is only subsequently held in place by adhesions due to minor local tissue reactions. This technique is therefore associated with a higher incidence of electrode displacements as compared with the open surgical implantation, which for this reason retains its value.

Wire electrodes for percutaneous insertion have been used in various forms: one single-lead electrode acting as a cathode in a monopolar arrangement, the device case serving as the anode; two single-lead electrodes in a bipolar arrangement; and one multiple-lead electrode in a bipolar arrangement with many combination possibilities. The latter electrode type for multilevel trials over the upper cervical segment, together with the necessity of extending the used frequencies from 100 to 1,400 Hz, was recommended by Waltz *et al.* for the treatment of dystonic and dyskinetic disorders (Waltz and Andreesen 1981, Waltz et al. 1981). The early implementation had been an assembly composed of 4 disc electrodes in linear array, spaced 1 cm apart, molded in Dacron mesh and silicone rubber, connected to a plug housing, thus requiring a partial laminectomy of C 4 to be inserted and percutaneous testing prior to permanent receiver implantation. A newer 4-electrode catheter assembly can be installed through an epidural needle and is connected with a computerized receiver which allows the lead combinations to be monitored on a continual basis through an external transmitter (Waltz 1982, Waltz and Davis 1983). In spite of the higher energy requirements as compared with cardiac pacemakers, totally implantable and programmable systems for neurostimulation have been produced, employing lithium batteries that provide several years of service. The stimulator and electrode system can be interrogated for impedance and parameters in use. The patient's compliance for the totally implantable system was reported to be much better than for the radiofrequency-linked systems (with external parts), the inconvenience of the former being the dependence of an attending doctor for reprogramming (Davis and Gray 1981).

3. Surgical Technique of Percutaneous Lead Implantation

The patient is placed prone on an X-ray permeable operating table, with a pillow under his thorax or abdomen in accordance with the intended site of puncture. An X-ray image amplifier system is necessary, that will visualize the tip of the puncture needle and follow the lead on its course within the spinal canal. After disinfection and draping of the patient's back, local anesthetic (1% lidocain, procain or mepivicain) is injected into the skin and the soft parts overlying the interspinous space aimed at. For a thoracic electrode placement this may be T 11–T 12 or T 12–L 1; for a cervical placement T 2–T 3 or T 3–T 4.

Saving the thoracolumbar fascia, a short axial midline skin incision is made either at this stage of the intervention or later on, after successful lead installation.

An epidural 15- or 16-gauge needle with upward angulation and provided with a stylet, is then introduced between two spinous processes, through the already opened skin, through the fascia and interspinous ligament. While the needle meets more resistance if it is advanced all the way through the interspinous ligament, this course ensures an optimal position of the needle tip at the dorsal midline notch of the spinal canal.

When the needle tip enters the epidural space by perforating the ligamentum flavum, a sudden decrease in resistance is felt. Intermittent X-ray control with sagittal beam direction is helpful in maintaining the epidural needle in the midsagittal plane. Negative pressure within the epidural space is a final criterion of correct needle position; audible aspiration of air or of a drop of sterile saline in the open hub of the needle, or otherwise the unresisting injectability of a small amount of saline without aspirability of blood or CSF, indicate an epidural placement.

The very flexible coil-wire lead is straightened by inserting a less flexible wire stylet, the front tip of which can be gently bent. This makes the lead steerable by axial rotation, once it is advanced through the needle and maneuvered in cephalad direction through the epidural space (Fig. 1). Sideward rotations of the needle tip may also contribute to overcome initial obstructions the lead possibly encounters on its way up. The electrode is positioned over the dorsal aspect of the cord in such a way that cathodal test stimulation (*e.g.*, 0.5–5 volt, 100 Hz and 200 μs pulse width) generates a tingling sensation within the dystonic body parts. The anodal electrode thereby is applied to an indifferent skin area. After successful lead placement, the wire stylet and then the epidural needle are prudently retracted, leaving only the coil-wire lead at its target, which can be monitored by X-ray control or by continuous test stimulation (Fig. 2). By means of sutures and a silicone cuff or a plastic button, the lead is anchored at the thoracolumbar fascia. Subsequently it is plugged on a percutaneous extension.

For permanent implantation the lead is connected with a receiver or a totally implantable power-supply and control device. The latter is placed in a bluntly dissected subcutaneous pocket in the subclavicular or subcostal region, under local or general anesthesia (Fig. 3). The subcutaneous tunneling from the pocket to the midback incision, harboring the connection leads, can indeed cause severe discomfort to the patient, in spite of

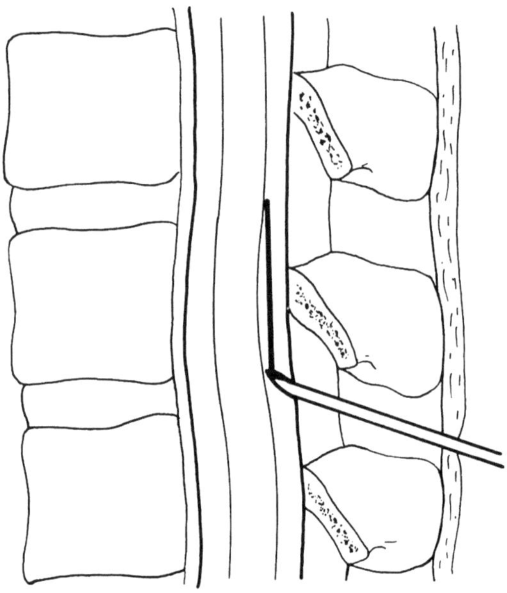

Fig. 1

ample local infiltration and the use of an appropriate tunneling instrument, so that for this stage of the intervention general anesthesia should be considered as an alternative, depending on the patient's individual tolerance. A teflon bag may be wrapped around the connection plugs, in order to avoid adhesions to the subcutaneous tissue, renderning easier eventual operative revisions later on. The wounds are closed in two layers.

For the sake of consolidation of the—in principle mobile—electrode on its very target, we impose confinement to bed and limit spinal mobility for a few days after the intervention. The percutaneous stimulation test usually lasts 7–10 days.

Minor variations from the above depicted technique are possible without affecting the essential principles (Ray 1981a, 1981b, Siegfried and Hood 1983).

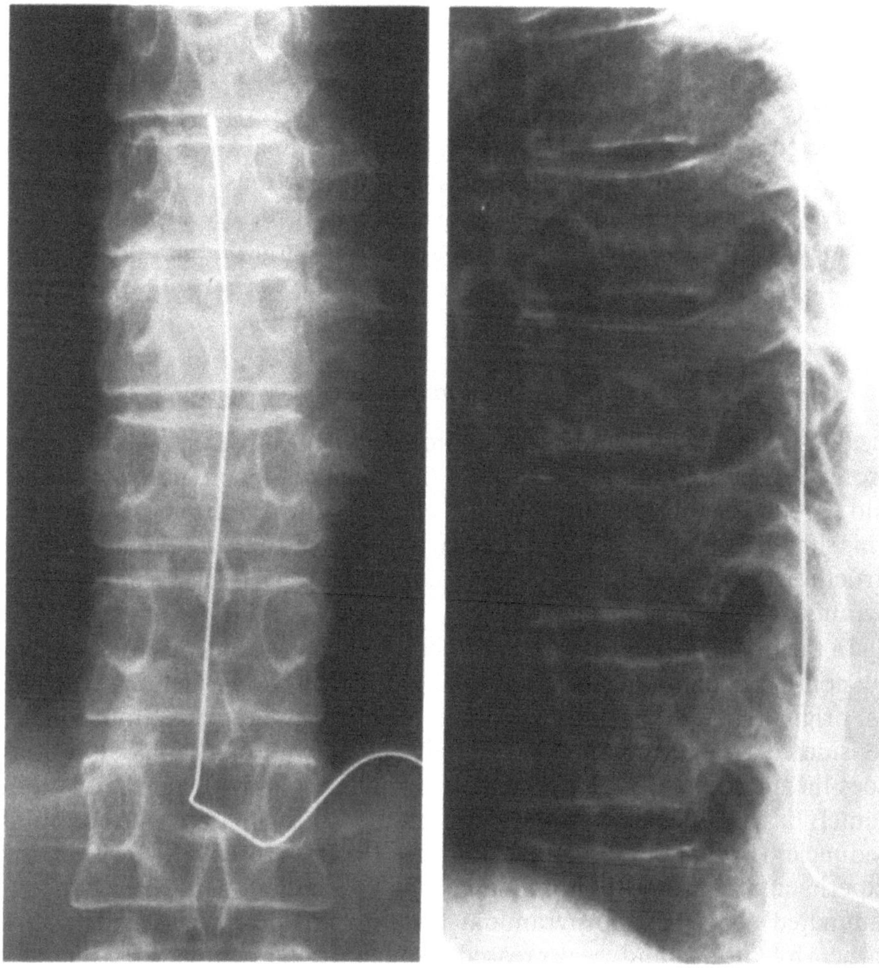

Fig. 2

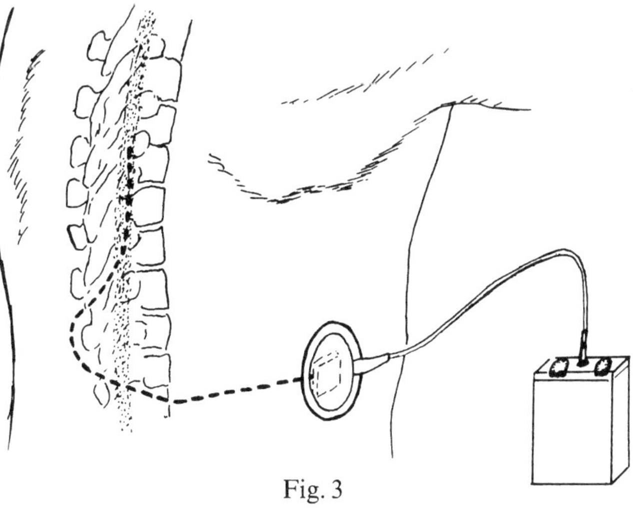

Fig. 3

II. Spasticity, a Synopsis

Spasticity evokes the image of rigidly stretched or, inversely, flexed and even contracted joints, of manual clumsiness and gait impairment, of flexor and extensor synergies, of hyperreflexia, clasp-knife reactions and clonus, and of an inversed plantar reflex (Babinski's sign). It is readily associated with hemiparesis, para- or tetraparesis, and with a lesion of the upper motor neuron, especially at the motor cortex, the internal capsule or the pyramidal tract.

1. Definition—Anatomical Considerations

Dealing with spasticity, the classic notion of the "pyramidal tract syndrome" arises, and this is used as another common denominator of spastic and paretic phenomena that causally refer to a single anatomical structure. This simplicity is not justified, neither with respect to any unique anatomical structure responsible for spasticity, nor with respect to the key role of pyramidal tract.

In the first place, the pyramidal tract is not a homogeneous fiber system but consists of components that differ anatomically and functionally, while its origin, as well as motor cortex, also covers regions traditionally considered as "extrapyramidal". In only one or two topographic situations does the monosynaptic component of the pyramidal tract run in isolation, namely at the bulbar pyramids and at the central part of the cerebral peduncle (Brodal 1975, Wiesendanger 1984). Moreover, there exists experimental evidence in subhuman primates, that lesions at those "uncontaminated" parts of the pyramidal tract do not produce spasticity, but rather hypotonia and a decreased though partially reversible manual dexterity (Wiesendanger *et al.* 1981, Kuypers 1982, Wiesendanger 1984). This has been confirmed in humans by the outcome of operations performed in order to relieve abnormal involuntary movements (Bucy *et al.* 1964).

Those findings do not contradict the general agreement that the main motor outflow from the cerebral cortex is contained in the pyramidal tract and that messages travelling along this tract are important for the skilled usage of the musculature (Allen and Tsukahara 1974) nor that this tract may contribute to the control of postural reactions and the "focusing" on somatosensory stimuli (Wiesendanger 1984). Nonpyramidal descending pathways also however, seem to play an important role in voluntary motor control, especially in cases with long-standing lesions of the pyramidal tract. The likelihood of interrupting all descending connections from the motor cortex and hence of provoking paresis and spasticity, is therefore greater with capsular than with more caudal lesions (Wiesendanger 1984).

A definition, recently formulated by Lance (Lance 1980), states: "Spasticity is a motor disorder characterized by a velocity-dependent increase in tonic stretch reflexes (muscle tone) with exaggerated tendon jerks resulting from hyperexcitability of the stretch reflex as one component of the upper motor neuron syndrome."

2. Experimental Approach

For more than a century, experiments have been undertaken in order to localize structures that are important in the generation of spasticity. Mesencephalic transection at intercollicular level, originally performed by Brondgeest in 1860, was repeated in the cat and described in detail by Sherrington in 1898. This "decerebration" produced a state of rigidity, most intense in antigravity muscles, including the erectors of the tail, the retractors and extensors of the neck.

Apart from some differences in detail and from the fact that "decerebrate" rigidity is present immediately after the mesencephalic transection, whereas poststroke spasticity develops slowly, the former nevertheless offered enough features in common with the latter to become a successful and convenient model for neurophysiological and pharmacological research (Wiesendanger 1985a).

As in the classical decerebrate preparation, spasticity in humans is abolished by dorsal root section. The injection of a small amount of procain into the muscle, insufficient to affect its response to motor stimulation, produces the same effect (Liljestrand and Magnus 1919, Walshe 1924). "Decerebrate" rigidity and human spasticity thus appear to be exaggerations of a proprioceptive reflex originating from a receptor organ within the muscle (spindle) and modulated by small motor fibers (gamma fibers). This aspect of spasticity thus received the designation "*gamma spasticity*".

As opposed to the classical mesencephalic transection preparation, an alternative anemic one, proposed in 1923 by Pollock and Davis and consisting of a ligature of the basilar and both carotid arteries, showed no abolition of rigidity by dorsal rhizotomy. This implies an isolated hyperexcitability of the alpha motoneurones, which gave rise to the designation "*alpha spasticity*" for this aspect of the disorder (Gros 1979).

As already mentioned, the decerebration rigidity model shows an acute onset, in contrast to the delayed onset of human spasticity following a spinal or supraspinal lesion. Interpreted in terms of the underlying pathophysiology, the model thus has a serious shortcoming.

Except for the spastic mutant mouse and rat, many other models have therefore been developed: by decortication, spinalization, temporary cord ischemia and deafferentation, revealing still other aspects of spasticity such as rhythmic automatisms, flexor spasms, "*interneuronal*" and "*deafferenta-*

tion" rigidity, respectively. Most realistic are the slowly developing conditions of increased muscle tone produced by spinal or cortical lesions in monkeys.

There is, however, no unique model for the clinical notion of spasticity, which has many different facets, depending mainly on the site of the lesion (Wiesendanger 1985a).

3. Physiological Considerations

Reversing the problem, this is not different from saying that the functionally normal neural machinery, in the execution of a voluntary movement, is called into operation at an array of hierarchic levels: 1. the motor unit; 2. spinal reflex mechanisms; 3. long spinal and brain stem control; 4. cerebellar vermis control; cerebrocerebellar controls 5. via the pars intermedia and 6. via the cerebellar hemispheres as well as via the basal ganglia (Eccles 1981).

1. The motoneuron with its muscle fibers forms the unitary basis of all movements, or otherwise, all movements are composites of contractions of individual units, as Sherrington recognized (Sherrington 1931). Another type of motoneuron, the small gamma-motoneuron, innervates the muscle spindles, which produce no appreciable tension, but activate proprioceptive sensory receptors. Phasic and static gamma-fibers, as well as combined skeletofusimotor neurones (beta-motoneurones) can be distinguished (Pierrot-Deseilligny and Mazières1984).

2. The monosynaptic spinal reflex is brought about by afferents (through I a-type fibers) from the stretch receptors in the muscle spindles to the alpha-motoneurones, so producing the tendon jerk or postural tone. Superimposed on this, is the gamma-motoneuron activation of the muscle spindles, heightening their sensitivity to stretch and so augmenting the monosynaptic responses (Fig. 4).

Conversely, these responses can be depressed by the disynaptic inhibitory pathway from the stretch receptors of the antagonist muscles (Fig. 5). The interplay of opposing muscle groups, not only for voluntary movements but also for the balance of tone, requires an interaction beyond the proper segment, that is: reciprocal innervation on an intersegmentary scale.

The receptors in the Golgi tendon organs have a higher threshold in their response to stretch, and act (through I b-type fibers) in general to excite flexor motoneurones and to depress extensors. Given a certain degree of stretch, they cause a brisk release of resistance (clasp-knife phenomenon).

Other high-threshold receptors of the muscle and cutaneous afferents have been termed "flexor reflex afferents" (FRA) by Lundberg in 1979. A role in locomotion (stepping) as ascribed to them. Instead of evoking a fixed pattern of ipsilateral flexion and crossed extension, the FRA dispose of

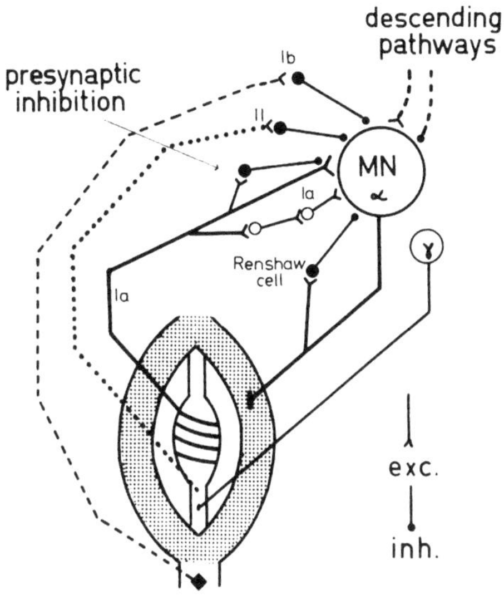

Fig. 4. Schematic wiring diagram showing some spinal reflex pathways. Excitatory synapses are represented by bars and inhibitory synapses by small black circles. Excitatory interneurones are represented by open circles and inhibitory interneurones by filled circles. (With permission from Pierrot-Deseilligny and Mazières, 1984)

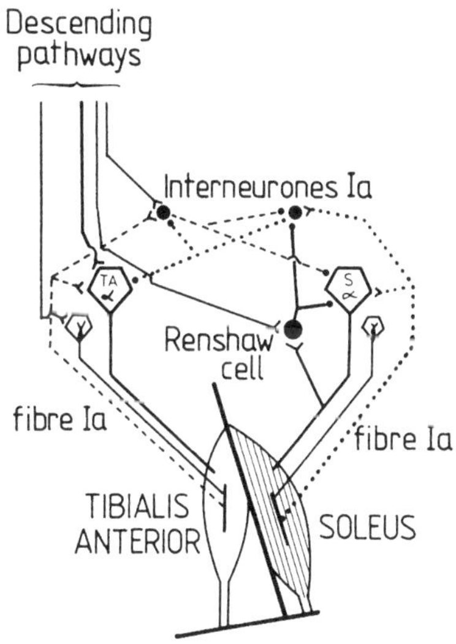

Fig. 5. Schematic diagram showing monosynaptic I a excitatory pathways from soleus (*S*) and tibialis anterior (*TA*) and the pathways of reciprocal Ia inhibition from *S* to *TA* and from *TA* to *S*. Same symbols as in Fig. 4 (excitatory and inhibitory synapses). Note that Ia interneurones parallely inhibit antagonistic MNs and opposite I a interneurones. Note that Renshaw cells activated by soleus MNs parallely inhibit soleus MNs and corresponding I a interneurones (for simplicity sake, Renshaw cells fed by TA MNs are not presented). (With permission from Pierrot-Deseilligny and Mazières 1984)

alternative excitatory and inhibitory reflex lines to both flexors and extensors, which are activated depending at least partially on the position of the limb (Lundberg 1979).

An important inhibitory mechanism is produced by the feedback control from motor axon collaterals via Renshaw cells, that inhibit motoneurones and interneurones, and that seem to be concerned in the adjustment of fine movements (Hultborn 1979 in Eccles 1981). Another important control is exerted by presynaptic inhibition, particularly effective in reducing the synaptic excitatory action of the primary afferent fibers (I a) entering the spinal cord via the dorsal roots (Eccles 1981) (Fig. 4).

3. Vestibular inputs provide supraspinal control of limb position by pathways from the vestibular nuclei down the vestibulospinal tracts, and from the reticular system of the brain stem down the reticulospinal tract (Wilson 1979 and Peterson 1979 in Eccles 1981).

4. A wide range of spinocerebellar pathways (anterior and posterior spinocerebellar tracts; cuneo-, spino-olivo- and reticulocerebellar tracts) convey from skin, muscle and joints, information that after a dynamic loop operation within the cerebellar vermis, is projected back to the spinal cord through the vestibulospinal and reticulospinal tracts (Eccles 1981).

5. The cerebellar pars intermedia and the contralateral cerebral motor cortex are intensively interconnected via pyramidal tract fibers, the ventrolateral thalamus, the red nucleus, the pontine nuclei and inferior olive, the large reticular and the interpositus nucleus (Eccles 1981).

6. Movements initiated from the cerebrum suppose impulses to be discharged from the pyramidal cells of the motor association cortex. These pass also to the contralateral cerebellar hemisphere via relays in the pontine nuclei and the inferior olive, and return after cerebellar computation to their cortical origin through the ventral lateral thalamus. Another loop links the association cortex with the motor cortex over the basal ganglia (Eccles 1981).

7. The association cortex is the first activated in willing an action, and thereby builds up excitatory paths communicating with other cortical regions, the cerebellar hemispheres and the basal ganglia, so setting up dynamic loop operations by reverberatory circuits. This process may be called "preprogramming". It ensures the input of motor memories (skills) stored in the association cortex, and controls rapid—ballistic—movements which are too fast for feedback control from the periphery (Eccles 1981).

4. Pathophysiological Approach

While the previously mentioned, spasticity-related structures and mechanism had mainly been disclosed by animal experiments, their value in human spasticity remained rather speculative, until techniques became available that permitted a noninvasive assessment of the various aspects of spasticity. The application of those techniques in man led to quantitative results, so that it is now possible to define normal value ranges.

The H-reflex, described by Hoffmann in 1918, is still used for neurophysiological analysis. Being a triceps surae contraction elicited by a submaximal electrical stimulation of the (sensory part of the) tibial nerve, it represents an electrically-induced equivalent of the monosynaptic ankle jerk, with this major difference that it short-circuits the neuromuscular spindles and thus is independent of their tone.

The shorter-latency muscle contraction elicited by a direct stimulation of motor axons, is called the M-response.

The F-wave is the late response which appears in a muscle after supramaximal stimulation of its motor nerve, due to the centrifugal discharge from motoneurones, initiated by an antidromic axonal volley.

The following summary of findings and interpretations in the application of neurophysiological assessment for spasticity was given by Delwaide (Delwaide 1986).

1. An alpha-motoneuron hyperexcitability, as a first statistically significant feature of spastic patients, has been detected by measurements of the ratio Hmax/Mmax, where Hmax is the maximum amplitude of the H-reflex and Mmax the maximum amplitude of the direct motor response. This ratio reflects the proportion of the motoneurones which can be activated reflexively. The F-wave amplitude, also reflecting the degree of excitability of the motoneurones, is higher on the diseased side of spastic hemiplegic patients.

2. *No* hyperactivity of the gamma-system, when assessed by single-I a-fiber recordings, was found to be present in spastic patients.

3. The inhibition of monosynaptic reflexes by vibratory muscle stimulation, reflecting the (I a-)presynaptic inhibition, is reduced in spastics. Factorial analysis showed that this reduction in presynaptic inhibition is the trouble that correlates best with spasticity (Hayat 1979).

By the "vibration paradox" is meant concomitant inhibition of phasic reflexes and a likewise vibration-induced slow tonic response to the motoneurones known as "tonic vibration reflex". The latter depends on progressive facilitation in polysynaptic pathways of spindle afferents. Both paradoxical phenomena are influenced differentially in spinal cord lesion or pyramidal hemiplegia (Desmedt 1983).

4. The (I a-)reciprocal inhibition, tested by conditioning a monosynaptic reflex with stimulation of a nerve coming from an antagonist, is reduced in spasticity.

5. *No* reduced activity of the Renshaw cells, which mediate the recurrent inhibition, was detected by the application of a sophisticated technique utilizing tandem H-reflexes, proposed by Pierrot-Deseilligny *et al.* (1976).

6. The study of polysynaptic reflexes, indirectly reflecting the function of the interneurones, appears to indicate that the inhibitory ones are less active in spasticity.

The four proven pathophysiological abnormalities mentioned above, contribute in variable proportions to produce clinical spasticity. In other words, the intensity of spasticity is not a function of only one of them. Nor is there a relationship between the individual pathophysiological profile and the etiology of spasticity in a given patient. The conclusion (Delwaide 1986) is that by directing treatment individually towards the apparently fortuitous pathophysiological profile of a spastic patient, one may optimize on a logical basis the selective employment of various antispastic treatments, which until now have often led to rather unpredictable results.

5. *Biochemical Approach*

While changes in electric charge (depolarization-hyperpolarization) constitute a rather nonspecific dimension of neuronal activity, synaptic transmission with a variety of transmitter substances, brings in the necessary differentiation that allows so many neuronal systems to coexist and still to interact in specific ways.

In a tabular form, some of the currently known (Delwaide *et al.* 1983, Lenman 1985) spinal neurotransmitters are listed below.

It is clear that pharmacological access to synaptic transmission in spasticity, as in other disease, may be most interesting for selective sections within a vast complexity of neuronal interactions.

Table 1. *Spinal Neurotransmitters Intervening in Motor Function*

Motorneurones—neuromuscular junction	acetylcholine
Primary afferents	glutamate
Nociceptive fine-caliber afferents	substance P
Substantia gelatinosa	enkephalin
Cutaneous afferents	somatostatin, angiotensin cholecystokinin
I a-reciprocal inhibition	glycine
I a-presynaptic inhibition	GABA
Renshaw recurrent inhibition	glycine, acetylcholine
Corticospinal fibers	glutamate
Reticulospinal fibers	acetylcholine
Bulbo- and pontospinal	serotonin norepinephrine

6. Spastic Bladder Dysfunction

While spasticity involves the striated skelett muscle, it has no direct relationship with the smooth muscle. The pelvic floor and the muscular perineum, both consisting of striated muscle, can therefore become spastic. Spasticity-related bladder dysfunction comprises mainly a relaxation deficit in m. levatorani and m. sphincter urethrae during voiding. Moreover flexor reflexes of the legs, as well as afferents from the vesical, urethral and rectal mucosa, influence the pelvic sphincters.

A dyssynergia between detrusor activation and sphincter relaxation may further complicate the situation (Pedersen 1981).

III. The Place of Spinal Cord Stimulation Among Other Treatments of Spasticity

The Tables 2–4 show the different kinds of treatment of spasticity in present and former times. They are classified respectively as ablative, physical and electro-stimulation procedures, as well as drug therapies.

The earliest and, the same time, longest list is the one of ablative procedures. Although their results are not always permanent, the procedures themselves are. This irreversibility can be considered as a disadvantage, because it eliminates the possibility of a trial run. Hence, the definition of clear-cut indications for ablative procedures is of utmost importance.

From a physiological point of view, it is certainly of interest that favorable distant effects on upper limbs and speech have been observed after ablative procedures on the lumbosacral rootlets (Gros *et al.* 1967, Fraioli and Guidetti 1977). Modern types of neurotomy and rhizotomy, including results, are discussed in detail by Gros 1979, Broggi 1981, Sindou *et al.* 1985, and Siegfried and Lazorthes 1985.

Within the nonablative group, physiotherapy founded on proprioceptive muscular facilitation and patterning, deserves a special mention. In many cases with residual motor function, it alone is capable of restoring to a remarkable extent the usefulness of weak and/or spastic limbs. Promotors of other types of spasticity treatment often grant that their methods only create a better starting point for revalidation and are of no use if not combined or follwed by physiotherapy.

The possibility of fine does adjustments—in fact a feature of reversibility—is offered by both electro-stimulation procedures and drug therapies. While most drugs have use-limiting side effects after systemic administration in spasticity-reducing doses, more selective (*i.e.*, intrathecal) ways of administration have been proposed, which especially in relation to baclofen, seem very promising.

Table 2. *Ablative Procedures in the Treatment of Spasticity*

Ablative procedure	Initial publications	Indication	Action site	Comment
Tenotomy	Delpech 1823	Achilles tendon	joint	loss of support
Neurotomy	Lorenz 1897	obturator nerve	motor nerve output	functional loss
Neurotomy, partial at muscle entry	Stoffel 1912, Silfverskiold 1923, Phelps 1951	foot, locally in spastic limbs		temporary result (!)
Procain in muscle, in motor end plate	Walshe 1924, Matthews and Rushworth 1956		gamma drive, alpha + gamma drive	short-lasting
Alcohol into motor end plate	Tardieu 1962		gamma drive, alpha + gamma drive,	lasting 1–6 months without force reduction
Selective partial (4/5)	Gros 1979	tibial n. (f. popliteal) for spastic foot	motor output + sensory input	immediate, long-lasting intraop. electr. charting
neurotomy	Privat *et al.* 1981	obturator n.	motor output + sensory input	immediate, long-lasting intraop. electr. charting
Posterior rhizotomy	Foerster 1908		sensory input	"spastics into tabetics"
Anterior rhizotomy	Monro 1945, Tarlov 1966	"alpha-"spasticity	alpha drive	in case posterior rhizotomy fails
Chemical rhizotomy	Nathan 1959, Kelly and Gauthier-Smith 1959	minimal residual function	motor output + sensory input	
Selective (4/5) post. rhizotomy	Ouaknine 1965, Gros *et al.* 1967	L1–S1	sensory input	minimal secondary sensory and sphincter disorder; long-lasting; favorable distant effects on upper limbs + speech

Selective microsurgical post. rhizotomy	Sindou 1972	cervical/lumbosacral painful spastic flexion	nociceptive + myotatic afferents	intraop. electrophysiol.; tactile + proprioceptive sparing; long-lasting
Sectorial post. rhizotomy	Privat *et al.* 1976, Gros 1976	cervical/lumbar	afferences from muscles with handicapping, not with useful spasticity	intraop. charting
Functional post. rhizotomy	Fasano *et al.* 1976	cerebral palsy	afferent input	intraop. electrophysiol.
Partial (1/2) post. rhizotomy	Fraioli and Guidetti 1977		afferent input	75% long-lasting success
Post. cordotomy (dors. columns)	Feld and Pecker 1951			
Longitudinal front. myelotomy (lat. approach)	Bischof 1951	spastic paraplegia in flexion without resid. motor, bladder, sexual function	sensory-motor reflex arc + longitudinal interneurones	
T-shaped myelotomy (post. commissurotomy + front. griseotomy)	Pourpre 1960		sensory-motor reflex arc + spinothalamic afferences	spares pyramidal fiber terminations
Conus medullaris myelotomy	Laitinen and Singounas 1971		sensory-motor reflex arc + spinothalamic afferences	no bladder disorder
Pedunculotomy/ pyramidotomy	Walker 1949, Bucy *et al.* 1964	involuntary movements and spasticity	upper motor neuron upper motor neuron	flaccid palsy
Pulvinarolysis	Cooper *et al.* 1973	dystonia + pain	visual and auditory proprioception	brief improvement
Fastigiotomy (stereotactic)	Hassler and Riechert 1961	cerebral palsy truncular spasticity	supraspinal descending facilitation	small number of cases
Dentatotomy (stereotactic)	Heimburger and Whitlock 1965	cerebral palsy athetosis	cerebello-thalamic/ -rubro-spinal pathways	never complete or lasting effects
Cortex ablation	Fulton 1937, Klemme 1942	dystonia dystonia	upper motor neuron upper motor neuron	transitory decrease of spasticity
or topectomy	Bucy 1948	dyskinesia	upper motor neuron	

Table 3. *Physical and Electrostimulation Procedures in the Treatment of Spasticity*

Physical procedure	Initial publications	Indication	Action site	Comment
Muscle cooling	Knutsson and Mattsson 1969	before physiotherapy	gamma drive	
30–34 °C bath	Paeslack 1981	massive spinal spasticity	?	
Physiotherapy	Kabat 1953, Vojta 1984, Bobath and Bobath 1958	residual mot. function with abnormal pattern	proprioceptive muscular facilitation/patterning	
Electrostimulation procedure	**Initial publications**	**Indication**	**Action site**	**Comment**
Spast. muscle stim.	Duchenne 1855 Lee *et al.* 1950	spastic hemiparesis	?	delay of 1 hour
Antagonist stim.	Levine *et al.* 1952	spastic hemiparesis	?	
Subcutan. nerve stim. med., radial, saphen. n.	Walker 1982	clonus	(supra)segmental inhibition	max. after 1 hour
Radial n. stim. + median/ulnar n. phenolization	Kiwerski 1984	spastic hand	antagonist support	antogonist strengthening during temporary agonist inactivation
Spinal cord stim.	Cook and Weinstein 1973	to be discussed	?	
Cerebellar cortex stim. (ant./post.)	Cooper 1973	spast. + athetosis, young patients	Purkinje cells asc./desc. inhibition + antidromically?	functional gain better than spasticity; somatotopy
Dentate stim.	Schvarcz *et al.* 1979	cerebral palsy	cerebellar efferences	speech, posture, gait and balance also better
(Sub)thalamic stim. (z. incerta, VPL)	Mundinger and Neumüller 1982	cerebral spasticity	basal ganglia + asc. inhibition	

Table 4. *Drugs in the Treatment of Spasticity*

Drug	Initial publications	Indication	Action site	Comment
Dantrolene	Snyder *et al.* 1967	spasticity not relieved by cooling	release of Ca into muscle sarcoplasm	effect more on fast than on slow fibers; muscular weakness
d-tubocurarine, gallamine suxamethonium	Griffith and Johnson 1942, 1948	relaxation in anesthesia	neuromuscular junction	parenteral, short-acting; produces paralysis
Mephenesine, meprobamate	Berger and Bradly 1946, 1954	spasticity in paraplegia	polysynaptic reflexes	short-lasting longer activity
Propranolol	Mai and Pedersen 1976	clonus	?	little general antispastic effect
Diazepam	Paeslack 1965	spinal spasticity	presynaptic inhibition by GABA facilitation	recent variants for intrathecal administration
Tizanidine	Gonsette *et al.* 1984	hypertonus and clonus in MS	polysynaptic pathways presynaptic inhibition	antiparetic effect
Morphine, intrathecally	Struppler *et al.* 1983	spasticity + pain	nociceptive afferences on interneurones	
Baclofen, orally, intrathecally	Faigle and Keberle 1972 Penn and Kroin 1984	spinal spasticity, relieved by cooling	mono- and polysynaptic transmission	GABA-like substance 30–90 mg/day 5–50 µg/day over implantable drug delivery systems

A selective influence on spasticity-related structures has also been the aim of electro-stimulation. In this, the concept of selectivity may fit the stimulation of affected muscles, of the cerebellum and of extrapyramidal pathways; spinal cord stimulation, however, does not. Historically, it arose from a proven treatment for chronic pain and anatomically, it spreads inevitably to affect several constituents and areas of the spinal cord.

It goes without saying that treatment of spasticity is expected to diminish the disabling aspects of the disorder. Not so well-accepted is the notion of useful spasticity, in the sense that a spastic antigravity tonus may constitute the crutch the patient needs to stand and walk on. If therapy does not manage to transform spasticity into smoothly coordinated and forceful motivity, it must not take away this last bit of support either. This applies especially to ablative and drug therapy.

IV. Results of Spinal Cord Stimulation in Spasticity

1. Experimental Results

Experimental investigations about the claimed depressant effect of spinal stimulation on spasticity have been carried out on decerebrate cats (Chapman *et al.* 1983), as well as on primate models of spasticity (Wiesendanger *et al.* 1985).

In decerebrate cats, and in one decerebrate monkey, spinal stimulation was found to depress the tonic responses of the stretch reflex, and this effect often persisted several minutes after cessation of stimulation. Less prominently, however, and rarely outlasting the period of stimulation, the phasic responses of the stretch reflex werde also depressed by spinal stimulation.

An important observation was the critical position of the stimulating electrode, now and then responsible for an adverse, spasticity-augmenting effect.

In monkeys, the most satisfactory model of chronic spasticity was produced by a ventrolateral funicular lesion of the cord. Here, spinal stimulation achieved a 49($\pm$ 26.7)% reduction of biceps stretch reflex responses, especially of the early components, in 6 out of 8 test sessions, as well as a 53% decrease of the triceps background activity. After cessation of stimulation, however, a rebound increase of both biceps (118%) and triceps (236%) responses was observed.

Electrophysiological and histological evidence was furnished of deleterious effects from daily stimulations at relatively high intensities (up to 600 μA).The massive cord lesions thus produced, were difficult to differentiate from electrolytic lesions. On the contrary, reports exist of no local or distant alterations of the nervous tissue, but of huge dural scarring reactions after 8 months of SCS in dogs (Broseta *et al.* 1986).

Wiesendanger *et al.* (1985) conclude—on the basis of very limited experimental data—that in view of capricious results and possible noxious effects, they cannot recommend the clinical use of prolonged epidural stimulation of the dorsal cord surface.

2. Clinical Results

a) Previous Review

In 1982 we reviewed the literature on the results of SCS for the modification of dystonic and hyperkinetic conditions (Gybels and Van Roost 1985). The effect of SCS on spasticity was included in this study, scrutinizing 39 papers from 19 authors or teams, describing 1,008 treated patients[14, 18, 22, 26a, 27, 28, 38, 47, 53, 56, 59, 67, 90, 98, 99, 100, 102, 103, 108, 125, 126, 133].

We had difficulties in differentiating spasticity from other motor impairments and in filtering the respective results from the literature, because many times those concepts—undoubtedly interconnected—were used interchangeably. We thus experienced the advantage and the further need of result description in terms of reproductible tests, representing well-definded characteristics of motor function such as muscle strength, muscle tonus and coordination. Throughout the above-mentioned material about spinal stimulation, traditional clinical observations (eventually including a clinical disability rating) prevailed or were the only assessment in more than one third of the papers. Other, more technical, tests were used to various extents, as illustrated by Fig. 6.

In order to augment the validity of bare figures derived from papers that varied in meticulousness of assessment and numbers of patients, we matched the percentually expressed averages of each improvement category (no—fair—good—very good) for just the most critically analyzed series, just the largest series and for all series (see Table 5). Hereby we considered a series to be critically elaborated if it presented data from at least four tests in addition to a clinical examination (see also Fig. 6). Series containing at least 30 patients were considered large.

"No" stands for no improvement at all, or deterioration; "fair" (F) for a slight improvement that does not change the preexisting level of functioning and which often can only be perceived by neurological assessment; "good" (G) for an improvement that raises the preexisting level of functioning in a way that it is useful in daily life activities, and can be perceived by untrained eyes; "very good" (VG) for an improvement that leads to an spectacular gain of function.

The "overall" evaluation ranges from global functional amelioration irrespective of spasticity, motor or bladder function, to improvement of otherwise use, mentioned impairments as, for instance, torticollis.

In Table 5, the figures for spasticity and especially those for bladder function improvement are noteworthy because of the high percentages in

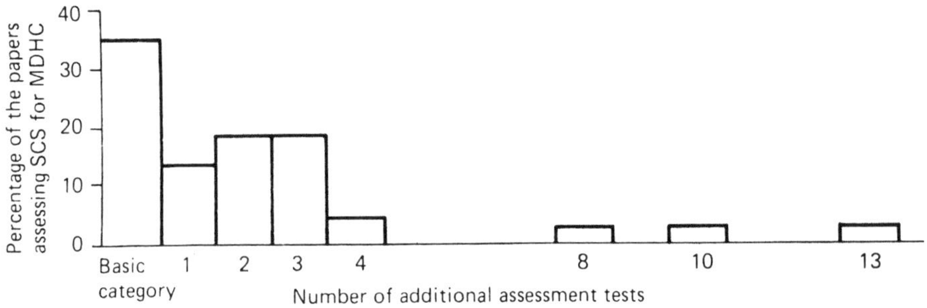

Fig. 6. Percentage of papers assesssing SCS for modification of dystonic and hyperkinetic conditions (MDHC) versus the number of additional assessment tests performed. The basic category consists of standard clinical examination. (From Gybels and Van Roost 1985, with courtesy of the publisher)

Table 5. *Mean Improvement, Expressed as Percentages of Numbers of Patients, Derived from the Most Critically Analyzed Series Only, from All Series, and from the Largest Series Only. (Material 1973–1982)*

	Overall				Motor				Spasticity				Bladder			
	No	F	G	VG	No	F	G	VG	No	F	G	VG	No	F	G	VG
Critical	60	12	18	10	68	16	16	0	44	35	0	21	36	17	21	26
Total	38	15	30	17	49	14	20	17	36	24	25	15	35	17	20	28
Large	40	14	26	19	44	18	20	18	39	21	22	18	37	25	12	27

the "very good" column, even when the critically assessed series alone are considered.

Within each improvement category, we then searched for statistical agreement among the "critical", "large" and "total" populations, and found it significant only for bladder function (X^2 test, contingency tables: $\alpha = 0.10$).

We think that this statistical analysis rules out a false positive result due to data compilation artifacts as far as the bladder is concerned. For spasticity, better figures than for other motor impairment, as compared with bladder function, may be statistically retrievable ($\alpha = 0.05$ in the Mann-Whitney test), but convincing unanimity fails.

Studying the material of 1973-1982, we further did not find sufficient comparable material to perform a numerical analysis about the alleged superiority of cervical over thoracic cord stimulation, or about the need for stimulation trials at high frequency.

Using not too stringent standards, we concluded that with SCS a useful relief of disability in terms of bladder function was obtained in 40–50% of the cases, a useful spasticity amelioration in maximally 20–40% of the cases, and a useful "overall" functional gain in 10 up to 45% of the cases.

b) Recent Results

In Table 6 we listed the data from 13 papers, out of which 12 were published since 1982 and one, published before 1982, was not taken into consideration in our previous review. Only one out of those 13 papers describes a series of more than 30 patients, which is an obvious shrinkage compared with the 7 "large" out of 22 series we gathered from 1973 to 1982. Apart from the shorter data collecting time recent investigators in this field might have had, a damped enthusiasm or a more critical look upon SCS could be the reason. Indeed, we observe a parallel increase of critical analysis. Four out of the recent 12 papers present data based on 4 technical tests in addition to clinical evaluation, against 4 others which belong to the basic category of Fig. 6. And even among the latter there is a tendency toward more standardization through the use of established rating scales.

In Table 6 a total number of 275 SCS-treated patients is represented, out of which only 63 in 8 papers were assessed on spasticity. In view of this relative paucity of material, we purposely omitted a statistical analysis this time.

A coarse comparison with Table 5, that displays the averages of the previously compiled material, tallies as far as the scores for motor performance and spasticity are concerned. The figures of bladder improvement, instead, seem to be shifted away from the "very good" rank. The corresponding authors moreover emphasize the poorly objectifiable (Berg *et al.* 1982) and/or long-lasting (Hawkes *et al.* 1983, Illis *et al.* 1983) benefit of SCS in the treatment of bladder function impairment (in MS). One team concludes that they cannot recommend it for routine management (Hawkes *et al.* 1983).

Concerning spasticity, it is striking that equally well analyzed series of patients show completely different outcome spectra (Barolat-Romana *et al.* 1985, Gottlieb *et al.* 1985). Curiously enough authors who treated spasticity due to spinal injury by means of stimulation below the lesion, had the best results (Barolat-Romana *et al.* 1985, Barolat-Romana and Myklebust 1986).

Waltz and Davis (1983) say that the site of SCS in alleviating spasticity is not critical, but Barolat-Romana *et al.* (1985)—in line with some previous reports (Richardson and McLone 1978, Richardson *et al.* 1979)—claim that in case of spinal transection it is compulsory below the lesion. The same authors also describe their best cases as responding immediately to

Table 6. *Improvement of Motor Performance, Spasticity and Bladder Function, Resulting from Spinal Cord Stimulation*

Authors pathology treated	Stimulation parameters	Motor performance				Spasticity				Bladder				Additional tests
		No	F	G	VG	No	F	G	VG	No	F	G	VG	
Barolat-Romana *et al.* 1985 cervical injury	upper thoracic 75–100 Hz 100–250 µs several hours/day					— (n=6)	—	3 50	3 50%					4
Barolat-Romana and Myklebust 1986 trauma and nontraumatic						2 (n=16) 12.5	—	6 37.5	8 50%					
Berg *et al.* 1982[a] MS, stabilized	C2–Th9 bladder: 15–35 Hz spast: 50–60 Hz	5 (n=11) 46	2 18	2 18	2 18%					2 (n=9) 22	3 33	3 33	1 11%	4
Broseta *et al.* 1986 spasticity	C2–C3 200–1,400 Hz 50–100 µs 38 months follow-up					4 (n = 5) 80	1 20%	—	—					1
dystonia		2 (n=3) 67	1 33%	—	—									
torticollis		1 (n=2) 50	—	—	1 50%									
Cioni and Meglio 1986[b] CVA	cervical 80 Hz 200 µs	3 (n=11) 27	1 9	7 64%	— —	2 (n = 10) 20	8 80%	— —	— —					4
Feeney and Gold 1980 MS	Th4; Th9 50 Hz 1 out of 4 hrs	— (n=1)	—	—	1 100%	—	1 100%	—	—	—	—	1 100%	—	1
Gottlieb *et al.* 1985 cerebral palsy	C2–C4 1–1,400 Hz 8–23 months stimul.					1	5	— (n=6)	—					3

Hawkes *et al.* 1983[c]	(Th 1–Th 9)									5	5–10	—	—	4
stable MS	(33 Hz)									(n = 15)				
	4–5 years follow-up									33	67%			
Illis *et al.* 1983[d]										7	1	4	—	3
MS	> 1 year follow-up									(n = 12)				
										58	8	33%	—	
Koulousakis *et al.* 1986	cervical/lumbar	16	—	—	—	4	4	7	1	5	3	4	—	
MS/myelopathy	resp. for tetra-	(n = 16)				(n = 16)				(n = 12)				
	and paraparesis	100%	—	—	—	25	25	44	6%	42	25	33%	—	
Nakamura and Tsubokawa	C 5; C 7					—	1	2	—					1
1985 sequel of apoplexia[e]	100–350 Hz					(n = 3)								
	300–500 µs					—	33	67%	—					
	12–14 hours/day													
Tallis *et al.* 1983	upper or mid-	16	4	2	1									
stable MS	thoracic	(n = 23)												
	33 Hz	70	17	9	4%									
	200 µs													
Waltz and Davis 1983	C 2–C 4	13	11	37	29									
cerebral palsy	100–1400 Hz	(n = 90)												
	6–64 months follow-up	15	12	41	32%									
dystonia	previous	17	4	17	17									
	thalamic surgery	(n = 55)												
	in 37/55 patients	31	7	31	31%									

Comments:

[a] With increasing pulse length, the tingling sensation is moving downward. Stimulation level C 2–C 4 strinkingly improves upper limb ataxia. Carry-over effect of SCS can last more than 30 days.

[b] Especially efficacious during voluntary movements and gait.

[c] Optimum response on day 6 or 7 of SCS. Only 3 out of 13 patients had lasting benefit. Complications in 9 out of 31 patients.

[d] In 3/4 of the patients the beneficial effect is lost at about 1 year, not only because of technical failure.

[e] Effect of SCS is delayed for 3–9 days; continuous improvement may be obtained for at least 4 months after the beginning of stimulation.

switching on/off the stimulator, whereby the favorable effect does not outlast the stimulator's activity. This is in striking contrast with other reports that mention considerable delays in response, and ameliorations outlasting the stimulation period.

3. About the Working Mechanism

At all events, the observed immediate effect of SCS below a complete cord transection (Barolat-Romana *et al.* 1985, Richardson and McLone 1978) indicates that its working mechanism may be purely spinal (propriospinal), and that here synaptic transmission rather than a slow biochemical process is involved. The latter and/or a functional and anatomical reorganization through plastic properties of the synaptic zone have been invoked (Illis 1983, Illis *et al.* 1976) in order to explain the usually observed delays between the beginning of stimulation and its functional effects (Dooley and Sharkey 1981, Hawkes *et al.* 1983, Illis 1983, Ketelaer *et al.* 1979, Nakamura and Tsubokawa 1985) or also between the withdrawal of SCS and the decay of effect (Berg *et al.* 1982, Cook 1976, Illis *et al.* 1980). Suprasegmental influence (through the ascending/descending reticular system (Waltz *et al.* 1981) have been invoked to explain effects on distant spinal segments or even on speech (Waltz and Davis 1983) and mood (Levita *et al.* 1981).

Actually, some 30 different working mechanisms of SCS have been postulated, but all remain more or less speculative. Except for the above-mentioned hypotheses, two others deserve special mention. The one of an involvement of the autonomic nervous system in reducing sympathetic tone (Read *et al.* 1980, Richardson *et al.* 1979, Dooley and Sharkey 1981) is based on documented skin temperature increases and evidence of healing arteriosclerotic ulcers. Since the occurrence of failure of objectifiable improvement with SCS, a placebo effect has also been suggested (Rosen and Barsoum 1979). This viewpoint has again been challenged (Illis *et al* 1980, Walker 1982), also by ourselves. We estimated from our statistical analysis that a major role of the placebo effect could be ruled out. Gottlieb *et al.* (1985) however, without deducing a placebo effect nominally, collected data that do not support the efficacy of cervical SCS for spasticity from an elegant experiment that once more furnishes food for thought. Seven patients were selected from 53 with spasticity treated by SCS. Only patients with definite changes in motor function as judged by themselves or their physicians, were considered. In a blind manner, clinical neurological and electrophysiological assessment was then requested to determine whether or not stimulation was on. While the patients were aware of their randomly predetermined stimulation status with nearly complete accuracy, the investigators did no better than chance! Moreover, in not one case did the clinical and electrophysiological tests support simultaneously correct judgment.

4. Complications

Complications of SCS that have repeatedly been reported, are: 1. lead breakage, eventually with current leakage, in some 10% of the cases; 2. electrode displacement in 9,4% of the cases and 3. infection in 4,7% of the cases. The two first complications are largely responsible for the fact that SCS is frequently not a single and simple, but a time-consuming and expensive technique (Illis 1983).

An exacerbation of MS within the first two weeks of stimulation has been reported in 5 out of 31 patients by Hawkes *et al.* (1981).

Other, more anecdotally reported complications were: pain at the stimulation site, CSF leakage, difficulties in keeping the electrodes buried under the skin, transient cord compression, epidural hematoma after laminectomy, hematoma at the receiver pocket, psychological difficulties in adjusting to the treatment program, a fibrotic reaction of the meninges presenting as pain or as reduction in stimulation intensity at a given voltage, and a reversible aggravation of spasticity by excessive stimulation.

5. Concluding Remarks

The appreciation of the value of spinal cord stimulation in the treatment of spasticity remains difficult. Our own attitude has been defined by the review we achieved in 1982, which included personal experience (Ketelaer *et al.* 1979). We estimated that SCS for spasticity has a questionable clinical value, a viewpoint we feel confirmed in by the recent results, though it certainly will not be shared by everyone.

Preliminary reports show that the intrathecal administration of minute doses of baclofen has very impressive effects on spasticity. Within the scope of limited personal experience with programmed continuous intrathecal baclofen delivery, we were able to observe an overwhelming medium-term effect on spasticity, which by its very obviousness only emphasizes the difficulty of appreciating the results of SCS in spasticity.

References

1. Allen GI, Tsukahara N (1974) Cerebrocerebellar communication systems. Physiol Rev 54: 957–1006
2. Barolat-Romana G, Myklebust JB (1986) Immediate effects of spinal cord stimulation on spinal spasticity. Presented at 7th Meeting ESSFN, Birmingham
3. Barolat-Romana G, Myklebust JB, Hemmy DC, Myklebust B, Wenninger W (1985) Immediate effects of spinal cord stimulation in spinal spasticity. J Neurosurg 62: 558–562

4. Berg V, Bergmann S, Hovdal H, Hunstad N, Johnsen HJ, Levin L, Sjaastad O (1982) The value of dorsal column stimulation in multiple sclerosis. Scand J Rehabil Med 14 (4): 183–191
5. Berger FM, Bradley W (1946) Pharmacological properties of α,β-dihydroxy-γ-(2-methylphenoxy)-propane (myanesin). Br J Pharamacol 1: 265–272
6. Bischof W (1951) Die longitudinale Myelotomie. Zbl Neurochir 11: 79
7. Bobath K, Bobath B (1958) An assessment of children with cerebral palsy and of their response to treatment. Am J Occup Ther 21: 19–31
8. Brodal A (1975) Neurological anatomy in relation to clinical medicine. 2nd ed. Oxford University Press, New York London Toronto
9. Broggi G (1981) Neurosurgical treatment of spasticity. In: Bauer HJ, Koella WP, Struppler A (eds) Therapie der Spastik. Verlag für angewandte Wissenschaften, München
10. Broseta J, Garcia-March G, Sánchez-Ledesma MJ, Barberá J, González-Darder J (1986) High-frequency cervical spinal cord stimulation in spasticity and motor disorders. 7th Meeting ESSFN 1986. Acta Neurochir [Suppl] 39: 112–116, 1987
11. Bucy PC (1948) Cortical exstirpation in the treatment of involuntary movements. Am J Surg 75: 257–263
12. Bucy PC, Keplinger JE, Siqueira EB (1964) Destruction of the "pyramidal tract" in man. J Neurosurg 21: 385–398
13. Campbell JN (1981) Examination of possible mechanism by which stimulation of the spinal cord in man relieves pain. Appl Neurophysiol 44: 181–186
14. Campos RJ, Dimitrijevic MM, Faganel J, Sharkey PC (1981) Clinical evaluation of the effect of spinal cord stimulation on motor performance in patients with upper motor neuron lesions. Appl Neurophysiol 44: 141–151
15. Chapman CE, Rüegg DG, Wiesendanger M (1983) Effects of dorsal cord stimulation on stretch reflexes. Brain Res 258: 211–215
16. Cioni B, Meglio M (1986) Spinal cord stimulation improves motor performances in hemiplegics: clinical and neurophysiological study. 7th Meeting ESSFN 1986. Acta Neurochir [Suppl] 39: 103–105, 1987
17. Cook AW (1976) Electrical stimulation in multiple sclerosis. Hosp Practice 11: 51–58
18. Cook AW, Taylor JK, Nidzgorski F (1981) Results of spinal cord stimulation in multiple sclerosis. Appl Neurophysiol 44: 55–61
19. Cook AW, Weinstein SP (1973) Chronik dorsal column stimulation in multiple sclerosis. NY St J Med 73: 2868–2872
20. Cooper IS, Amin I, Chandra R (1973) A surgical investigation of the clinical physiology of the LP pulvinar complex in man. J Neurol Sci 18: 89–110
21. Davis R, Gray E (1981) Technical factors important to dorsal column stimulation. Appl Neurophysiol 44: 160–170
22. Davis R, Gray E, Kudzma J (1981) Beneficial augmentation following dorsal column stimulation in some neurological diseases. Appl Neurophysiol 44: 37–49
23. Delpech J (1823) Ténotomie du tendon d'Achille. Chirurgie clinique Montpellier ou observations et réflexions tirées de travaux de chirurgie clinique de cette école. Paris-Gabon

24. Delwaide PJ (1986) Spasticity: from pathophysiology to therapie. Lecture at 7th Meeting ESSFN, Birmingham
25. Delwaide PJ, Schoenen J, Burton L (1983) Central actions of neurotropic drugs assessed by reflex studies in man. In: Desmedt JE (ed) Motor control mechanism in health and disease. Raven Press, New York
26. Desmedt JE (1983) Mechanism of vibration-induced inhibition or potentiation: tonic vibration reflex and vibration paradox in man. In: Desmedt JE (ed) Motor control mechanisms in health and disease. Raven Press, New York
26a. Dooley DM (1977) Neurosurgery 1: 220–224
27. Dooley DM, Nisonson I (1981) Treatment of patients with degenerative diseases of the central nervous system by electrical stimulation of the spinal cord. Appl Neurophysiol 44: 71–76
28. Dooley DM, Sharkey J (1981) Electrical stimulation of the spinal cord in patients with demyelinating and degenerative diseases of the central nervous system. Appl Neurophysiol 44: 218–224
29. Duchenne GB (1855) De l'électrisation localisée et de son application à la pathologie et à la thérapeutique, 1 vol. JB Baillière, Paris
30. Eccles JC (1981) Physiology of motor control in man. Appl Neurophysiol 44: 5–15
31. Faigle JW, Keberle H (1972) The metabolism and pharmacokinetics of Lioresal. In: Birkmayer W (ed) Spasticity. A topical survey. Hans Huber, Bern Stuttgart Vienna
32. Fasano VA, Barolat-Romana G, Ivaldi A, Sguazzi A (1976) La radicotomie postérieure fonctionelle dans le traitement de la spasticité cérébrale. Neuro-Chirurgie 22: 23–24
33. Feeney DM, Gold GN (1980) Chronic dorsal column stimulation: effects on H reflex and symptoms in a patient with multiple sclerosis. Neurosurg 6: 564–566
34. Feld M, Pecker J (1951) Essai de cordotomie postérieure dans la paraplégie spasmodique; Application au syndrome paraplégique de Little. Communic Soc Neurochirurgie de Langue Francaise. Séance de 4–5 déc
35. Foerster O (1908) Über eine neue operative Methode der Behandlung spastischer Lähmungen mittels Resektion hinterer Rückenmarkswurzeln. Z Orthop Chir 22: 203
36. Fraioli B, Guidetti B (1977) Posterior partial rootlet section in the treatment of spasticity. J Neurosurg 46: 618–626
37. Fulton JF (1937) Spasticity and the frontal lobes. A review. N Engl J Med 217: 1017–1024
38. Gildenberg PL (1981) Comprehensive management of spasmodic torticollis. Appl Neurophysiol 44: 233–243
39. Gonsette RE, De Smet Y, Demonty L (1984) Tizanidine, a new antispastic agent: 6 years clinical experience in 152 MS patients. In: Gonsette RE, Delmotte P (eds) Immunological and clinical aspects of multiple sclerosis. Proceedings of the XXV anniversary symposium of the Belgian Research Group for Multiple Sclerosis. MTP Press Ltd, Lancaster Boston, The Hague, Dordrecht
40. Gottlieb GL, Myklebust BM, Stefoski D, Groth K, Kroin J, Penn RD (1985) Evaluation of cervical stimulation for chronic treatment of spasticity.

Neurology 35: 699–704
40a. Griffith HR, Johnson GE (1942) Anesthesiology 3: 418–420
41. Gros C (1979) Spasticity—Clinical classification and surgical treatment. In: Krayenbühl H et al (eds) Advances and technical standards in neurosurgery, vol 6. Springer, Wien New York, pp 55–97
42. Gros C, Quaknine G, Vlahovitch B, Frerebeau P (1967) La radicotomie sélective postérieure dans le traitement neurochirugical de l'hypertonie pyramidale. Neuro-Chirurgie 13: 505–518
43. Gybels J, Van Roost D (1985) Spinal cord stimulation for the modification of dystonic and hyperkinetic conditions. A critical review. In: Eccles JC, Dimitrijevic MR (eds) Recent achievements in restorative neurology—1 Upper motor neuron functions and dysfunctions. Karger, Basel
44. Hassler R, Riechert T (1961) Wirkungen der Reizungen und Koagulationen in den Stammganglien bei stereotaktischen Hirnoperationen. Nervenarzt 32: 97–108
45. Hayat A (1979) Factorial analysis of specific parameters in spasticity. Electromyogr Clin Neurophysiol 19: 541–553
46. Hawkes CH, Beard R, Fawcett D, Paul EA, Thomas DGT (1983) Dorsal column stimulation in multiple sclerosis; effects on bladder and long term findings. Br Med J 287 (6395): 793–795
47. Hawkes CH, Fawcett D, Cooke ED, Emson PC, Paul EA, Bowcock SA (1981) Dorsal column stimulation in multiple sclerosis: effects on bladder, e.g. blood flow and peptides. Appl Neurophysiol 44: 62–70
48. Heimburger RF, Whitlok CC (1965) Stereotaxic destruction of the human dentate nucleus. Confin Neurol 26: 346–358
49. Hoffmann P (1918) Über die Beziehungen der Sehnenreflexe zur willkürlichen Bewegung und zum Tonus. Z Biol 68: 351
50. Illis LS (1983) Central nervous stimulation in neurological disease. J Roy Soc Med 76: 905–909
51. Illis LS, Read DJ, Sedgwick EM, Tallis RC (1983) Spinal cord stimulation in the United Kingdom. J Neurol Neurosurg Psychiat 46: 299–304
52. Illis LS, Sedgwick EM, Oygar AE, Sabbahi Awadalla MA (1976) Dorsal column stimulation in the rehabilitation of patients with multiple sclerosis. Lancet June 26: 1383–1386
53. Illis LS, Sedgwick EM, Tallis RC (1980) Spinal cord stimulation in multiple sclerosis: clinical results. J Neurol Neurosurg Psychiat 43: 1–14
54. Kabat H (1953) Proprioceptive facilitation techniques for treatment of paralysis. Phys Therapy Rev 33–2
55. Kelly RE, Gautier-Smith PC (1959) Intrathecal phenol in the treatment of reflex spasm and spasticity. Lancet ii: 1102–1105
56. Ketelaer P, Swartenboeckx G, Deltenre P, Carton H, Gybels J (1979) Percutaneous epidural dorsal cord stimulation in multiple sclerosis. Acta Neurochir (Wien) 49: 95–101
57. Kiwerski J (1984) New possibilities of improving the function of the hand of patients with spastic hemiplegia. Int J Rehab Research 7: 293–298
58. Klemme RH (1942) Surgical treatment of dystonia. With report of one hundred cases. Research Publications. Ass. for Research in Nervous and Mental Diseases 21: 596–601

59. Klinger D, Kepplinger B (1982) Quantification of the effects to epidural spinal electrostimulation (ESES) in central motor disorders. Appl Neurophysiol 45: 221–224
60. Knutsson E, Mattsson E (1969) Effects of local cooling on monosynaptic reflexes in man. Scand J Rehabil Med 1: 126–132
61. Koulousakis A, Buchhaas U, Nittner K (1986) Application of SCS for movement disorders and spasticity. 7th Meeting ESSFN 1986. Acta Neurochir [Suppl] 39: 112–116, 1987
62. Kuypers HGJM (1982) A new look at the organization of the motor system. Progr Brains Res 57: 381–403
63. Laitinen LV, Singounas E (1971) Longitudinal myelotomy in the treatment of spasticity of the legs. J Neurosurg 35: 536–540
64. Lance JW (1980) Symposium Synopsis. In: Feldmann RG, Young RR, Koella WP (eds) Spasticity: disordered motor control. Chicago, Year Book. pp 485–494
65. Larson SJ, Sances A Jr, Riegel DH, Meyer GA, Dallmann DE, Swiontek T: Neurophysiological effects of dorsal column stimulation in man and monkey. J Neurosurg 41: 217–223
66. Law JD (1983) Spinal stimulation: Statistical superiority of monophasic stimulation of narrowly separated longitudinal bipoles having rostral cathodes. Appl Neurophysiol 46 (1–4): 129–137
67. Lazorthes Y, Siegfried J, Broggi G (1981) Electrical spinal cord stimulation for spastic motor disorders in demyelinating diseases—a cooperative study. In: Hosobuchi Y, Corbin I (eds) Indications for spinal cord stimulation. Excerpta Medica, Princeton
68. Lee WJ, McGovern JP, Duval EN (1950) Continuous tetanization (low voltage) currents for relief of spasm. Arch Phys Med 31: 766–770
69. Lenman JAR (1985) The pharmacology of spasticity. In: Eccles JC, Dimitrijevic MR (eds) Recent achievements in restorative neurology—1 Upper motor neuron functions and dysfunctions. Karger, Basel
70. Levine MG, Knott M, Kabat H (1952) Relaxation of spasticity by electrical stimulation of antagonist muscles. Arch Phys Med 33: 668–673
71. Levita E, Rilan M, Waltz JM (1981) Psychological effects of spinal cord stimulation: preliminary findings. Appl Neurophysiol 44: 93–96
72. Liljestrand G, Magnus R (1919) Pflügers Arch ges Physiol 176: 168
73. Long DM, Erickson D, Campbell J, North R (1981) Electrical stimulation of the spinal cord and peripheral nerves for pain control. A 10-year experience. Appl Neurophysiol 44: 207–217
74. Lorenz (1897) Über die chirurgische Behandlung der angeborenen spastischen Gliedstarre. Wien Klin Rdsch 21: 25–27
75. Lundberg A (1979) Multisensory control of spinal reflex pathways. In: Granit R, Pompeiano O (eds) Reflex control of posture and movement. Prog Brain Res 50: 11–28
76. Mai J, Pedersen E (1976) Clonus depression by propranolol. Acta Neurol Scand 53: 395–398
77. Matthews PCB, Rushworth G (1956) Differential nerve narcosis with procaine. J Physiol (London) 131: 30 P

78. Melzack R, Wall PD (1965) Pain mechanism: a new theory. Science 150: 971–979
79. Monro D (1945) The rehabilitation of patients totally paralysed below waist: anterior rhizotomy for spastic paraplegia. N Engl J Med 233: 456–461
80. Mundinger F, Neumüller H (1982) Programmed stimulation for control of chronic pain and motor disease. Appl Neurophysiol 45: 102–111
81. Nakamura S, Tsubokawa T (1985) Evaluation of spinal cord stimulation for postapoplectic spastic hemiplegia. Neurosurg 17 (2): 253–259
82. Nathan PW (1959) Intrathecal phenol to relieve spasticity in paraplegia. Lancet ii: 1099–1102
83. Ouaknine G (1965) La radicotomie sélective postérieure dans le traitement de l'hypertonic pyramidale. Thèse Méd, Montpellier
84. Paeslack V (1981) Die Bedeutung spinaler Spastik bei der Rehabilitation des Paraplegikers. In: Bauer HJ, Koella WP, Struppler A (eds) Therapie der Spastik. Verlag für angewandte Wissenschaften, München
84a. Paeslack V (1965) Paraplegia 3: 91
85. Pedersen E (1981) Miktionsstörungen und Spastik. In: Bauer HJ, Koella WP, Struppler A (eds) Therapie der Spastik. Verlag für angewandte Wissenschaften, München
86. Penn RD, Kroin JS (1984) Intrathecal baclofen alleviates spinal cord spasticity. Lancet May 12: 1078
87. Phelps WM (1951) Treatment of paralytic disorders exclusive of poliomyelitis. In: Bancroft FW, Marble HC (eds) Surgical treatment of the motor-skeletal system. Lippincott Co, Philadelphia
88. Pierrot-Deseilligny E, Bussel B, Held JP, Katz R (1976) Excitability of human motoneurones after discharges in a conditioning reflex. Electroenceph Clin Neurophysiol 40: 279–287
89. Pierrot-Deseilligny E, Mazières L (1984) Circuits réflexes de la moelle épinière chez l'homme. Contrôle au cours mouvement en rôle fonctionell. 1re et 2e partie. Rev Neurol 140 (11): 605–614, (12): 681–694. Masson SA Edit, Paris
90. Plotkin R (1980) Pisces stimulation for motor neurone disease. Acta Neurochir (Wien) 30 [Suppl]: 429–433
91. Pollock LJ, Davis LE (1923) Studies in decerebration. I. A method of decerebration. Arch Neurol Psychiatry (Chicago) 10: 391–398
92. Pourpre MH (1960) Traitement neurochirurgical des contractures chez les paraplégiques post-traumatiques. Neuro-Chirurgie 6: 229–236
93. Privat JM, Benezech J, Frerebeau P, Gros C (1976) Sectorial posterior rhizotomie: a new technique of surgical treatment of spasticity. Acta Neurochir (Wien) 35: 181–195
94. Privat JM, Privat C, Frerebeau P, Benezech J, Gros C (1981) Renouveau de la neurotomie obturatrice. In: Simon R. (ed) Acutalités en rééductation fonctionelle et réadaptation. 6e série. Masson, Paris, pp 70–76
95. Ranck JB Jr (1975) Which elements are excited in electrical stimulation of mammalian central nervous system: a review. Brain Res 98: 417–440
96. Ray CD (1981a) Percutaneous spinal cord stimulation: technique. In: Hosobuchi Y, Corbin T (eds) Indications for spinal cord stimulation. Proceedings of a symposium conducted at 7th International Congress of Neurological Surgery. Excerpta Medica, München

97. Ray CD (1981a) Spinal epidural electrical stimulation for pain control. Practical details and results. Appl Neurophysiol 44: 194–206
98. Read DJ, Matthews WB, Higson RH (1980) The effect of spinal cord stimulation on function in patients with multiple sclerosis. Brain 103: 803–833
99. Reynolds AF, Oakley JC (1982) High frequency cervical epidural stimulation for spasticity. Appl Neurophysiol 45: 93–97
100. Richardson RR, Cerullo LJ, McLone DG, Gutierrez FA, Lewis V (1979) Percutaneous epidural neurostimulation in modulation of paraplegic spasticity. Six case reports. Acta Neurochir (Wien) 49: 235–243
101. Richardson RR, McLone DG (1978) Percutaneous epidural neurostimulation for paraplegic spasticity. Surg Neurol 9: 153–155
102. Rosen JA, Barsoum AH (1979) Failure of chronic dorsal column stimulation in multiple sclerosis. Ann Neurol 6: 66–67
103. Scerrati M, Onofrj J, Pola P (1982) Effects of spinal cord stimulation on spasticity. H-reflex study. Appl Neurophysiol 45: 62–67
104. Schvarcz JR, Sica RS, Morita E (1979) Estimulación crónica del núcleo dentado en espasticidad por paràlisis cerebral. Boln Asoc Argent Neurocir 26: 74
105. Sherrington CS (1898) Decerebrate rigidity and reflex coordination of movements. J Phiol 22: 319–332
106. Sherrington CS (1931) Quantitative management of contraction in lowest level coordination. Brain 54: 1–28
107. Sherwood AM (1981) Biomedical engineering aspects of spinal cord stimulation. Appl Neurophysiol 44: 126–132
108. Siegfried J (1980) Treatment of spasticity by dorsal cord stimulation. Int Rehab Med 2: 31–34
109. Siegfried J, Hood T (1983) Current status of functional neurosurgery. In: Krayenbühl H et al (eds) Advances and technical standards in neurosurgery, vol 10. Springer, Wien New York
110. Siegfried J, Lazorthes Y *et al* (eds) (1985) Neurochirurgie de l'infirmité motrice cérébrale. Neuro-Chirurgie 31 [Suppl 1], Masson, Paris
111. Silfverskiold N (1923) Reduction of the uncrossed two-joint muscles of the leg to one-joint muscles in spastic condition. Acta Chir Scand 56: 315–322
112. Sindou M (1972) Étude de la jonction radiculo-médullaire postéricure. La radicellotomie postérieure sélective dans la chirurgia de la douleur. MD Thesis nr. 173, Lyon
113. Sindou M, Pregelj R, Boisson D, Eyssette M, Goutelle A (1985) Surgical selective lesions of nerve fibers and myelotomies for modifying muscle hypertonia. In: Eccles JC, Dimitrijevic MR (eds) Recent achievements in restorative neurology—1 Upper motor neuron functions and dysfunctions. Karger, Basel
114. Shealy CN, Mortimer JT, Reswick JB (1967) Electrical inhibition of pain by stimulation on the dorsal columns; preliminary clinical report. Anesth Analg (Cleve) 46: 489–491
115. Snyder HR jr, Davis CS, Bickerton RK, Halliday RP (1967) 1-((5-Arylfurfurylidene)amino)hydantoins. A new class of muscle relaxants. J Med Pharm Chem 10: 807–810

116. Stoffel A (1912) The treatment of spastic contracture. Am J Orthop Surg 10: 611
117. Struppler A, Burgmayer B, Ochs GD, Pfeiffer HG (1983) The effect of epidural application of opoids on spasticity of spinal origin. Life Sciences 33 [Suppl] 1: 607–610
118. Tallis RC, Illis LS, Sedgwick EM (1983) The quantitative assessment of the influence of spinal cord stimulation on motor function in patients with multiple sclerosis. Int Rehab Med 5: 10–16
119. Tardieu G (1962) Action des infiltrations d'alcool dilué sur diverses raideurs liées aux lésions cérébrales. Bull et Mém Soc Méd Hôp de Paris 113: 1, 7, 12
120. Tarlov IM (1966) Deafferentation to relieve spasticity or rigidity: reasons for failure in some cases of paraplegia. J Neurosurg 25: 270–274
121. Vojta V (1984) Die cerebralen Bewegungsstörungen im Säuglingsalter. F Enke, Stuttgart
122. Walker AE (1949) Cerebral pedunculotomy for the relief of involuntary movements and hemiballismus. Acta Psychiat Kbh 24: 723–729
123. Walker JB (1982) Modulation of spasticity: prolonged suppression of a spinal reflex by electrical stimulation. Science 216: 203–204
124. Walshe FMR (1924) Procaine injection for spasticity. Brain 47: 159–172
125. Waltz JM (1982) Computerized percutaneous multi-level spinal cord stimulation in motor disorders. Appl Neurophysiol 45: 73–92
126. Waltz JM, Andreesen WH (1981) Mutiple-lead spinal cord stimulation: technique. Appl Neurophysiol 44: 30–36
127. Waltz JM, Davis JA (1983) Cervical cord stimulation in the treatment of athetosis and dystonia. Adv Neurol 37: 225–237
128. Waltz JM, Reynolds LO, Riklan M (1981) Multi-lead spinal cord stimulation for control of motor disorders. Appl Neurophysiol 44: 244–257
129. Wiesendanger M (1985a) Animal models of motor disorders. In: Struppler A, Weindl A (eds) Electromyography and evoked potentials. Springer, Berlin Heidelberg New York
130. Wiesendanger M (1985b) Is there an animal model of spasticity? In: Delwaide PJ, Young RR (eds) Clinical neurophysiology in spasticity. Elsevier Science Publ, Amsterdam
131. Wiesendanger M (1984) Pyramidal tract function and the clinical "pyramidal syndrome". Human Neurobiol 2 (4): 227–234
132. Wiesendanger M, Chapman CE, Rüegg DG (1981) Welche Rolle spielt die Pyramidenbahn in der Genese der Spasticität? In: Bauer HJ, Koella WP, Struppler A (eds) Therapie der Spastik. Verlag für angewandte Wissenschaften, München, pp 87–98
133. Young RF, Goodmann SJ (1979) Dorsal spinal cord stimulation in the treatment of multiple sclerosis. Neurosurgery 5: 225–230

B. Technical Standards

Dorsal Root Entry Zone (DREZ) Thermocoagulation

D. G. T. THOMAS

Department of Neurological Surgery, Institute of Neurology, The National Hospital, London (U.K.)

With 9 Figures

Contents

Introduction

Dorsal root entry zone thermocoagulation, that is DREZ lesioning, is useful in the treatment of pain associated with deafferentation. The procedure can control, in many cases, pain caused by avulsion of the brachial plexus, by herpes zoster infection of the spinal dorsal root ganglia, and pain due to traumatic paraplegia. All these conditions are difficult to control by other means including drug treatment, or by surgical methods like sympathectomy, rhizotomy, cordotomy or neurostimulation. The precise anatomical and physiological basis for the success of the procedure is not known, but it is probably due to destruction of pain generating centers within the spinal cord as well as to a readjustment of inhibitory and excitatory inputs and an interruption of ascending and descending local reflex pathways.

Physiological and Anatomical Basis for DREZ Lesions

The DREZ consists anatomically of the central part of the dorsal spinal roots, Lissauer's tract and the superficial layers of the dorsal horn, where afferent fibers synapse with the cells of the spino-thalamic tract. Sindou (1974 a) demonstrated that at the point of entry of the posterior rootlet into the spinal cord the nociceptive small myelinated $A\delta$ and the unmyelinated C fibers lie laterally. These fibers enter the medial part of Lissauer's tract, which lies adjacent to the posterior lateral part of the dorsal horn. They may reach the posterior horn itself at the segmental level where they enter, or through an ascending or descending pathway up to two spinal segmental levels above or below. The dorsal horn itself can be divided on the basis of its cytoarchitecture into laminae, the Rexed laminations (1952), and the nociceptive afferents relay on lamina I or lamina IV and V. From here the ascending spinothalamic pathway ascends in the cord. Substance P, which can excite dorsal horn neurones responsive to noxious stimuli, is found in Rexed layers I, II, III, and IV (Blumenkopf 1984). The $A\beta$ fibers of the posterior rootlet lie medially on entry to the cord, and form long axons which ascend in the dorsal columns as well as short collateral fibers which enter the lamina II and III of the dorsal horn. Somatostatin, methionine or encephalin, all of which can inhibit nociception, may be found in layers I, II, IV and VII. Within the lateral part of the tract of Lissauer there are local intersegmental reflex pathways connecting different levels of the substantia gelatinosa. The anatomical and physiological basis of the gate control mechanism of pain (Melzack and Wall 1965) lies in the DREZ. Stimulation of the nociceptive afferents, which normally results in production of substance P at the dentrites of the spinothalamic cells, can be inhibited by stimulation of the large lemniscal fibers and their collaterals with production of metencephalin. Local segmental reflexes within Lissauer's tract inhibit or excite transmission of nociception, while descending fibers from the reticulospinal tract may also have further inhibitory effects. It is probable that when the large lemniscal primary afferents in the peripheral nerve or in the posterior nerve roots are lost, by disease or trauma, that there is an excessive firing in dorsal horn neurones causing deafferentation hyperactivity and the perception of deafferentation pain. DREZ lesions may reduce the nociceptive impulses generated in the spinothalamic pathways by destruction of the dorsal horn neurones together with the medial part of

Lissauer's tract. Sindou (1974 b) introduced surgical section of the lateral small fiber component of each posterior rootlet as it entered the DREZ as a method of selectively destroying the laterally placed small nociceptive fibers together with the medial, excitatory, part of Lissauer's tract in the treatment of carcinomatous infiltration of the brachial plexus, where the preservation of tactile and proprioceptive functions were particulary desirable. Nashold (1976) introduced a more extensive lesion of the DREZ using thermocoagulation of the posterior part of the dorsal horn, including the substantia gelatinosa and the medial part of the tract of Lissauer, in order to control deafferentation pain due to brachial plexus avulsion. It is the latter method which will be described further.

Indications and Patient Selection

The principal indications for DREZ thermocoagulation are neuropathic pain due to avulsion injuries of the nerve roots, postherpetic neuralgia and pain related to paraplegia. Results are less encouraging in deafferentation due to peripheral nerve injury, rhizotomy, thoracotomy and amputation.

a) Pain in Plexus Avulsion Injury

The anterior and posterior spinal roots of the brachial or lumbosacral plexuses may be completely or partially avulsed from the spinal cord by severe traction forces applied to the limbs and limb girdles in accidental trauma, particularly in road traffic accidents affecting motor cyclists. Wynn Parry (1984) has described the characteristic pain experienced by patients with avulsion of the brachial plexus. About 90% of patients with such injuries develop pain in the deafferentated limb. The pain generally starts within days or weeks on the injury, although the exact timing may be difficult to elicit if there is significant associated head injury. The pain is severe and usually burning or crushing in character. Generally it is constant and concentrated in one region of the damaged limb, although not in a distribution corresponding to a precise segmental root distribution. In many patients the pain becomes less spontaneously so that within about three years it remains a significant problem in about 25%. Noninvasive methods of pain treatment can be very useful in controlling the pain. Physical rehabilitation, and where possible the return to work, usually have a beneficial effect. Transcutaneous electrical stimulation, in the root of the neck above the area rendered anesthetic by the lesion and in the inner arm in a T 2 root distribution, can be very beneficial. Drug treatment with simple analgesics together with the anticonvulsant drug carbamezepine or sodium valproate may also be useful.

Amitriptyline, with both antidepressant and analgesic effects, also may be of use. Many patients report that alcohol, in moderate amounts, also lessen the pain greatly. However, there is a minority of patients, probably in the region of about 5% overall, who remain in desperately severe pain in spite of treatment. In some patients the pain appears to increase in severity even many years after the original injury. It is these patients who will be offered DREZ thermocoagulation as pain treatment.

b) Pain in Postherpetic Neuralgia

Acute herpes zoster infection of a dorsal root ganglion in a spinal nerve root is associated with a vesicular skin rash and acute pain in a dermatomal distribution. In most cases the pain gradually settles as the rash resolves, but in about 10% it persists (de Morgan and Kierland 1957). In elderly patients the proportion with lasting postherpetic neuralgia is probably disproportionately higher than in younger patients. The superficial component of the pain may be described as burning or aching and as being made worse by light touch in areas around the anesthetic, healed, scars. The second, deep, component is paroxysmal and gripping in nature, not associated with triggering by touch. Other unpleasant feelings in the affected part can include itching and a feeling of swelling or of constriction. Treatment in the acute attack with antiviral agents may reduce the ultimate incidence of postherpetic neuralgia. However, the established condition can be very resistant to treatment. Transcutaneous stimulation as well as drug treatment with carbamazepine (Gerson 1977), amitriptyline (Taub 1973, Watson 1982) or with chlorprothixene (Nathan 1978) may be successful. A minority of patients will be left with severe and disabling symptoms and in these DREZ thermocoagulation can be effective in spite of age and other infirmity (Friedman, 1984 a, b).

c) Chronic Pain in Traumatic Paraplegia

Following spinal cord injury patients may be left with pain at the site of the original injury due to mechanical factors like instability or facet joint disruption. However, there can also be pain of a central type experienced in the deafferentated region below the injury, often consisting of burning or tingling feelings in the trunk or anesthetic limbs (Melzack 1978, Nashold 1981 a). Pain triggered by touch and movement may also be experienced in the anesthetic areas, and this pain may be described as tingling or lancinating. DREZ thermocoagulation at, and for two to three segments above, the spinal level may be beneficial for the pain experienced in the deafferentated areas. As in the cases of avulsion, or postherpetic pain, the procedure is a serious one and should be considered only for severe intractable symptoms. A specific risk in the paraplegic cases is that the spinal level is likely to ascend, certainly in sensory terms and often in motor terms, to higher segments following the lesion.

Preoperative Investigation

Myelography

In cases of avulsion of the plexus or traumatic paraplegia patients have often undergone myelography soon after injury in order to define the extent of root and spinal cord damage. In the cases of brachial plexus avulsion, if myelography has not been performed previously, it is useful to perform this study as part of the preoperative assessment prior to DREZ thermocoagulation. Information is obtained from the myelogram not only about the

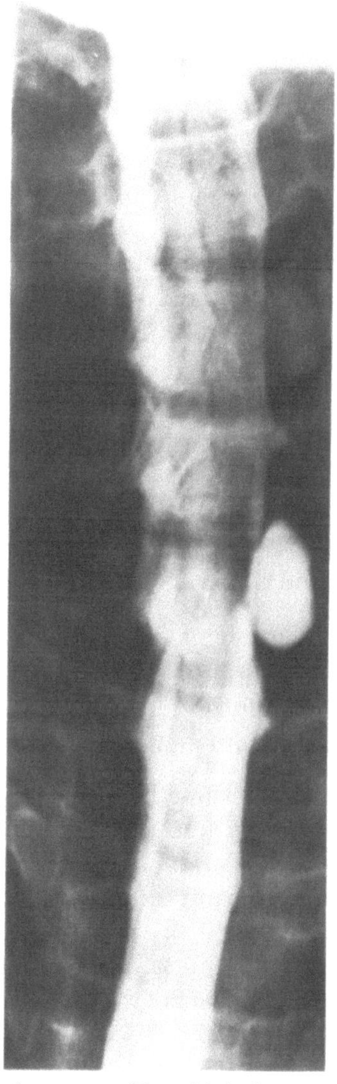

Fig. 1. Cervical myelogram in a case of brachial plexus avulsion. Absence of roots and a typical pseudomeningocoele visualized on injured side

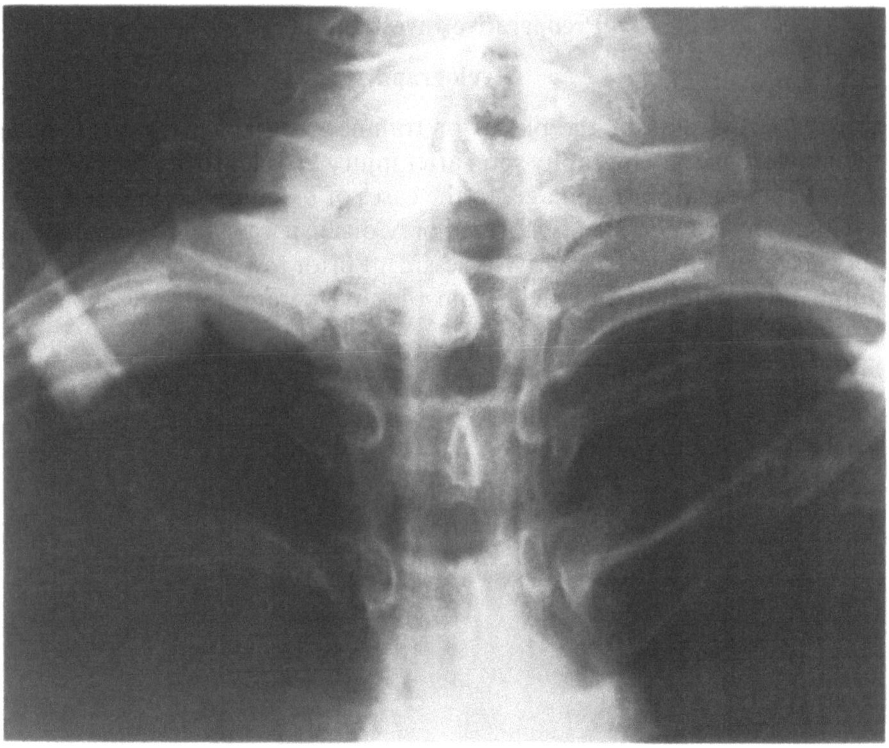

Fig. 2. Cervical myelogram in a case of brachial plexus avulsion. Very large pseudomeningocoele, extending to the clavicular region visualized

extent of root loss but also about the existence of traumatic meningoceles, which can be very large, and which may affect the surgical approach (Figs. 1 and 2).

Surgical Technique

Preparation and Positioning

24 hours prior to surgery dexamethazone 4 mg. 6 hourly is commenced together with cimetidine 200 mg 6 hourly. The patient is operated on under general anesthesia in a prone position. In cases of brachial plexus avulsion it is possible to determine the relevant spinal levels accurately by palpation of the prominent C 7 vertebra when the patient is positioned with the neck flexed and by subsequent palpation of the C 1 or C 2 vertebrae during the course of laminectomy. In cases where the dorsal region is involved marker X-rays performed preoperatively, together with X-ray control peroperatively, is generally essential. Identification of levels in the dorsal region and at the conus can be facilitated by the use of peroperative physiological methods, considered below.

Incision, Laminectomy and Dural Opening

In the case of brachial plexus avulsion midline cervical incision is made and full laminectomy from C 4-T 1 performed (Figs. 3–5). In the dorsal region incision and laminectomy is carried out two levels above the affected dermatome from D 1–D 6 and three levels above from D 6–D 10. The conus

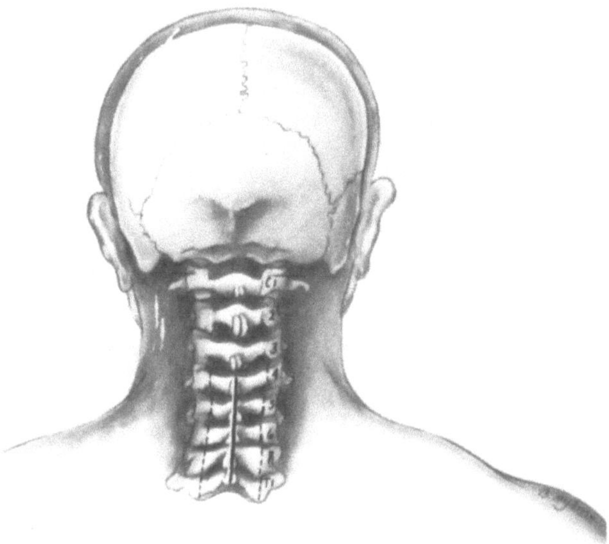

Fig. 3. Bony extent of laminectomy required in DREZ lesion for brachial plexus avulsion

is exposed through a laminectomy from D 10–L 1. Scrupulous hemostasis should be achieved prior to dural opening. A dispersive earth electrode, generally in the form of a 10 cm solid metal disposable lumbar puncture needle should be implanted in the superficial muscle form the earth electrode of the thermocoagulation system. The circumference of the dura is sealed with lintine and towels to exclude skin and muscle prior to dural opening. The dura is incised in the midline and held with stay sutures (Figs. 6 and 7). The operating microscope is used to provide both magnification (in the range 10–15 times) and illumination. The arachnoid,

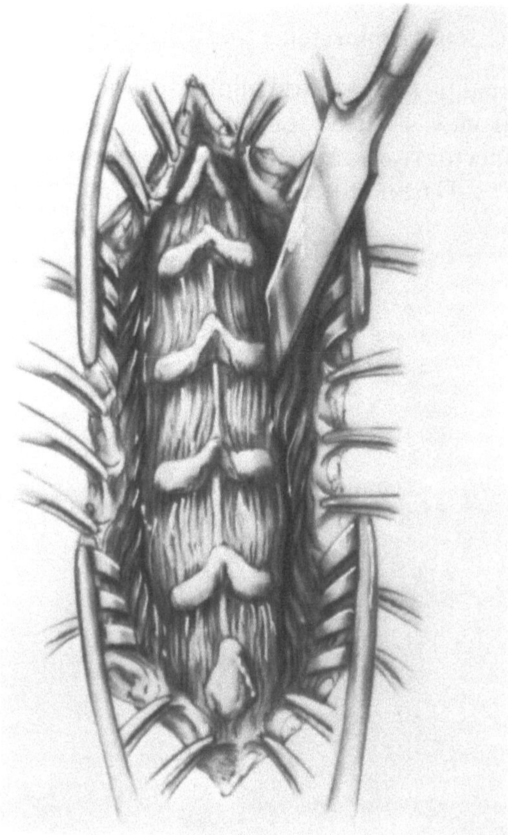

Fig. 4. Stripping of muscle layers to expose cervical laminae

which is frequently thickened and opaque in cases of trauma, as it is in some cases of postherpetic neuralgia, is opened to expose the spinal cord. In cases of root avulsion the DREZ is identified by inspection and palpation. The anatomy of the normal, uninjured side, together with the anatomy just above and below the affected levels, gives an indication of the probable site of the DREZ. Universally, in avulsion cases, there is a degree of hemi-atrophy of the cord on the injured side (Fig. 7). In addition, there may be displacement or rotation of the cord caused by arachnoid cysts or pseudomeningoceles. Visual inspection of the cord may reveal the sulcus between the posterior and lateral columns and this identification of the DREZ may be confirmed by gentle palpation with a metal dissector passed across the surface of the cord. In cases of postherpetic neuralgia the affected rootlets are often wasted and atrophic, as well as being affected particularly by arachnoiditis. In cases of traumatic paraplegia preoperative my-

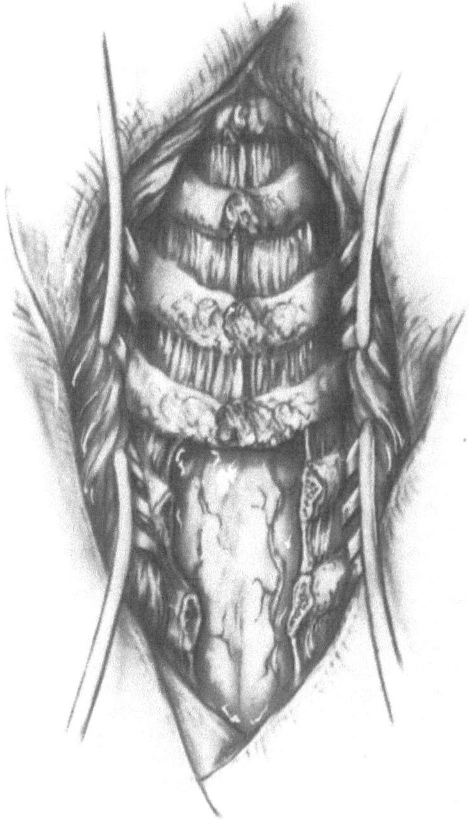

Fig. 5. Laminectomy of cervical vertebrae

elography, together with inspection at the time of surgery, shows the level of trauma. There is generally a degree of arachnoiditis at this level and above.

In the dorsal region, and in the conus, electrical stimulation may be particularly useful in identifying the appropriate root level (Friedman and Nashold 1984 a, b). In the region of the conus the last dentate ligament is at approximately the L 5 segmental level, while the S 1 root is approximately 1 cm above the point where the filum terminale joins the conus.

DREZ Thermocoagulation

Once the appropriate spinal levels have been identified and the sulcus between the posterior and lateral columns confidently defined, thermocoagulation of the DREZ may be carried out. Lesions are made using a specially constructed insulated electrode; containing a temperature sensitive

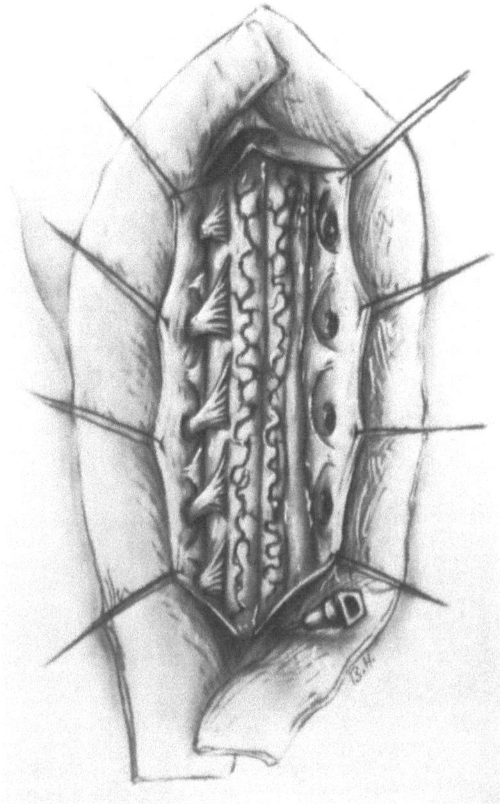

Fig. 6. Dural opening and exposure of spinal cord. Avulsed roots on injured side

thermocouple (NCTD thermocouple temperature monitoring electrode Radionics Inc. Burlington, Massachusetts). The electrode has an exposed tip of 2 millimeters and diameter of 0.25 millimeters. A circuit is made by connecting cables between this active electrode, and the radiofrequency generator (RFG-5 with thermocouple adaptor TCA-1; Radionics Inc). The earth electrode is as described above. In order to demonstrate the continuity of the circuit, impedance in the system is first tested, using the special facility available in the generator. Typically the impedance between the dura, or the spinal cord, and earth will be in the region of 450–700 ohms. It is necessary repeatedly to clean the electrode, as well as to clean the connecting plugs and sockets, in order to have the optimum conditions in the electrical circuit during thermocoagulation. Under vision through the microscope the electrode is introduced into the DREZ to a depth of 2 millimeters (Fig. 8). The electrode is hand-held and introduced at an angle of approximately 25° to the parasagittal plane. Radiofrequency current is switched on and

Fig. 7. Operative photograph in case of partial avulsion of brachial plexus. Several posterior roots remain partially intact. Dura retracted by stay sutures, dissector placed in pseudomeningocoele. There is relative rotation and displacement of the cord away from the side of injury

gradually increased until a temperature of 70–75 °C is reached. This temperature is held constant for 15 seconds and then switched off with removal of the electrode (Nashold 1984). The current setting is then returned to zero and the second, and subsequent, lesions made in a similar way, with withdrawal and cleaning of the electrode between lesions. Lesions are sited 2 millimeters apart. Occasionally bleeding will be encountered from small spinal pial vessels, but this will usually stop with pressure from a

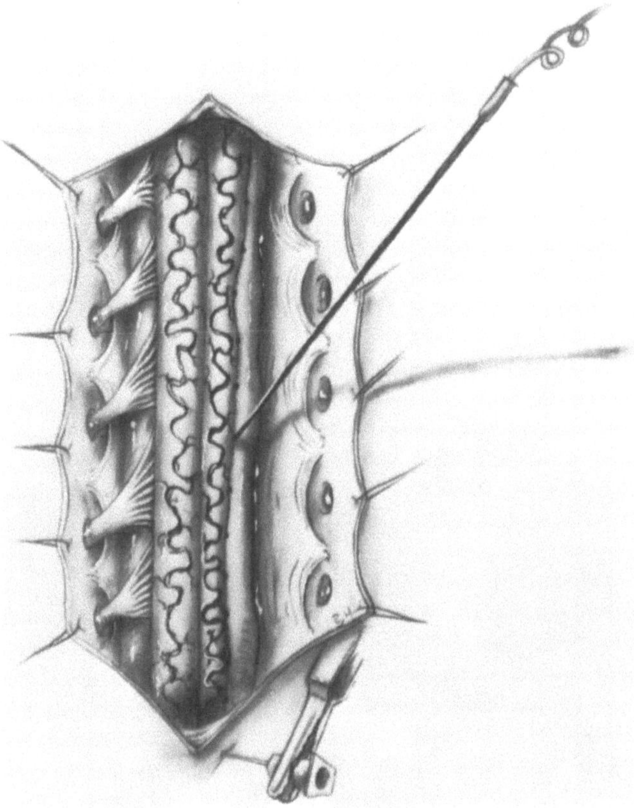

Fig. 8. DREZ exposed and hand held electrode being used for coagulation

pattie without requiring bipolar diathermy. In cases of brachial plexus avulsion approximately 20–24 lesions are required to cover the DREZ from C 5-T 1. In the dorsal region, for postherpetic neuralgia, the affected root level together with one segment above and one segment below is treated. In cases of paraplegia the level of the lesion and two segments above are treated. The radiofrequency current required to obtain a temperature of 70–75 °C in the cervical region is about 45 mA and generally in the dorsal region it is much less, usually about 25 mA. The most accurate method of obtaining consistent lesions is with the use of temperature measurement, as described above.

Closure

The dura is closed in a water tight fashion with routine closure of the muscle, subcutaneous layers and skin with interrupted sutures (Fig. 9).

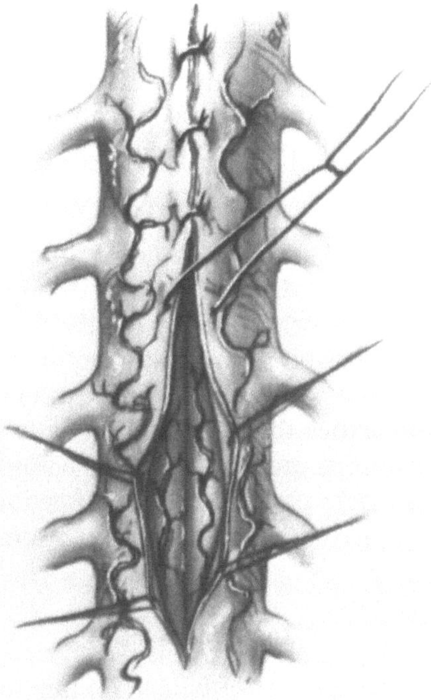

Fig. 9. Closure of dura

Peroperative Electrophysiological Methods

Electrical evoked potentials may be elicited after stimulation of the nerve roots as they pass through the spinal foramina, or by stimulation of the intercostal, posterior tibial, sural or sciatic nerves using percutaneous electrodes. Recording at the conus, or higher in the cord in postherpetic cases, of these evoked potentials, on the undamaged side, can contribute to determining the anatomy and appropriate segmental levels peroperatively (Friedman 1984a, b).

Preoperative studies of the cortical somatosensory evoked potentials in brachial plexus avulsion cases often shows subclinical spinal cord damage. However, peroperative use of electrophysical monitoring of spinal cord function has not yet been successful in identifying, and limiting, long tract complications from DREZ lesioning (Jones 1985).

Postoperative Care

The patient is nursed in bed, with rolling from side to side 4 to 6 hourly, for 48 to 72 hours before mobilising to stand with assistance. In some cases a urinary catheter must be passed if spontaneous micturition is not successful. In cases of cervical laminectomy a soft cervical collar is useful for increased comfort in the postoperative period. Intermittent lumbar punctures in order to reduce CSF pressure and so to lessen the risk of postoperative fistula may be employed. Providing there is no significant additional neurological deficit the steroid cover may be discontinued after a few days. In cases where there is significantly increased disability, due to long tract involvement, steroids should be continued and physiotherapy and rehabilitation instituted. Analgesia with opiates or with pethidine may be required to control wound pain for several days. Many of the patients with chronic pain undergoing DREZ lesions appear to have a relatively low threshold for pain and in the postoperative period it may take two or three days before they can distinguish their wound pain from the preexisting neuropathic pain. As the postoperative pain settles they will be gradually able to estimate the degree of pain relief from the procedure. In this period objective responses to pain relief will also be seen in the form of reduced consumption of analgesic drugs and improved behavior and function, compared with the preoperative state.

Complications

In approximately half the cases of DREZ lesioning for avulsion pain there will be an observable increase in signs of neurological deficits due to long tract involvement postoperatively (Thomas 1984). In most cases this will be a very minor change in motor function or reflexes, or in sensation. Often the deterioration in motor function is a clumsiness and ataxia of the ipsilateral lower limb, without objective pyramidal weakness on testing. Sometimes there is dysfunction of the posterior columns with impairment of proprioception. Generally, these changes are of little functional significance, and usually they tend to remit over a period of several weeks. However, in a minority of cases, less than 10%, there will be permanently persisting deficits due to the procedure. In a small proportion of these locomotion may be significantly limited, although in nearly every case of such complications in patients with avulsion or postherpetic neuralgia they remain able to walk unaided. The rate of such complications is lower in postherpetic than in posttraumatic avulsion cases, possibly because there is no spinal cord atrophy due to the original disease process.

Results

A good result, that is partial or complete resolution of pain, so that narcotic analgesic medication can be discontinued, is obtained in between two thirds and three quarters of cases of avulsion or postherpetic neuralgia (Friedman 1984 a, b, Levy 1983, Nashold 1979, Samii 1984, Thomas 1983, 1984) and in a rather smaller fraction of cases of pain due to traumatic paraplegia (Nashold 1981 a). Provided the pain relief is stable for a period of weeks postoperatively it appears to last indefinitely over a period of years. It is almost unheard of for a patient to find pain worsened by DREZ, but some patients who develop neurological complications may regret the procedure, even if they have obtained pain relief. However, for a significant majority of patients with deafferentation pain, of desperate severity, DREZ coagulation provided worthwhile pain relief.

References

1. Blumenkopf B (1984) Neuropharmacology of the dorsal root entry zone. Neurosurgery 15: 900–903
2. De Morgan JM, Kierland RR (1957) The outcome of patients with herpes zoster. AMA Arch Dermatol 75: 193–196
3. Friedman AH, Nashold BS Jr (1984 a) Dorsal root entry zone lesions for the treatment of postherpetic neuralgia. Neurosurgery 15: 969–970
4. Friedman AH, Nashold BS Jr, Ovelmen-Levitt J (1984 b) Dorsal root entry zone lesions for the treatment of post-herpetic neuralgia. J Neurosurg 60: 1258–1262
5. Gerson GR, Jones RB, Luscombe DK (1977) Studies on the concomitant use of carbamazepine and domipramine for relief of post herpetic plain. Postgrad Med J 53 [Suppl] 104–109
6. Jones SJ, Thomas DGT (1985) Assessment of long sensory tract conduction in patients undergoing dorsal root entry zone coagulation for pain relief. In: Schwann J, Jones SJ (eds) Spinal cord monitoring. Springer, Berlin Heidelberg New York, pp 266–273
7. Levy WJ, Nutkiewicz A, Ditmore QM, Watts C (1983) Laser-induced dorsal root entry zone lesions for pain control. J Neurosurg 59: 884–886
8. Melzack R, Wall PD (1965) Pain mechanism. A new theory. Science 150: 971–979
9. Melzack R, Loeser JD (1978) Phantom body pain in paraplegics: evidence for a central "pattern generating mechanism" for pain. Pain 4: 195–210
10. Nashold BS Jr, Urban B, Zorum DS (1976) Phantom pain relief by focal destruction of substantia gelatinosa of Rolando. Adv Pain Res 1: 959–963
11. Nashold BS Jr, Ostdahl RH (1979) Dorsal root entry zone lesions for pain relief. J Neurosurg 51: 59–69
12. Nashold BS Jr, Bullitt E (1981 a) Dorsal root entry zone lesions to control pain in paraplegics. J Neurosurg 55: 414–419
13. Nashold BS Jr (1981 b) Technical note: Modification of DREZ lesion technique. J Neurosurg 55: 1012
14. Nashold BS Jr (1984) Current status of the DREZ operation. Neurosurgery 15: 942–944
15. Nathan PW (1978) Chlorprothixene (taractan) in postherpetic neuralgia and other severe chronic pains. Pain 5: 367–371

16. Rexed B (1952) The cytoarchitectonic atlas of the spinal cord in the cat. J Comp Neurol 96: 415–495
17. Samii M, Moringlane JR (1984) Thermocoagulation of the dorsal root entry zone for the treatment of intractable pain. Neurosurgery 15: 953–955
18. Sindou M, Quoex C, Balleydier C (1974 a) Fiberorganization at the posterior spinal cord-rootlet junction in man. J Comp Neurol 153: 15–26
19. Sindou M, Fischer G, Goutelle A, Mansuy L (1974 b) La radicellotomie postérieure sélective. Premiers résultats dans la chirurgie de la doleur. Neurochirugie 20: 391–408
20. Taub A (1973) Relief of post-herpetic neuralgia with psychotropic drugs. J Neurosurg 39: 235–239
21. Thomas DGT, Sheehy JPR (1983) Dorsal root entry zone lesions (Nashold's procedure) for pain relief following brachial plexus avulsion. J Neurol Neurosurg Psychiatry 46: 924–928
22. Thomas DGT, Jones SJ (1984) Dorsal root entry zone lesions (Nashold's procedure) in brachial plexus avulsion. Neurosurgery 15: 966–967
23. Watson CP, Evans RJ, Reed K, Mershey H, Goldsmith L, Walsh J (1982) Amitryptiline versus placebo in post-herpetic neuralgia. Neurology (NY) 32: 671–673
24. Wynn Parry CB (1984) Pain in avulsion of the brachial plexus. Neurosurgery 15: 960–965

Acute Surgery for Ruptured Posterior Circulation Aneurysms

S. J. Peerless, S. Nemoto, and C. G. Drake

Division of Neurosurgery, The University of Western Ontario, London, Ontario, (Canada)

Contents

Introduction

The timing of surgery to obliterate ruptured intracranial aneurysms remains controversial. Although recent studies, which were uncontrolled, suggested a slight benefit to early surgery for aneurysms of the anterior circulation, the relative infrequency of aneurysms of the vertebral-basilar circulation has made the question of when to operate on these aneurysms even more uncertain[15]. Our experience with more than 1400 operated cases of aneurysms of the posterior circulation has made the surgical approach to the confining space in front of the brain stem quite routine. Nevertheless, most of the cases were operated upon 10 to 14 days following their last subarachnoid hemorrhage (SAH) and after the immediate effects of the hemorrhage and the consequences of the hemorrhage had passed. As well, there is little recorded evidence of an advantage or disadvantage of early surgery for these uncommon aneurysms[2, 12, 23, 27]. This report documents our experience with early operation in ruptured posterior circulation aneurysms during the past 15 years.

Clinical Material

Since 1970, 86 patients with ruptured posterior circulation aneurysms have been operated on at The University of Western Ontario within 7 days following their last SAH. These patients mean age was 47 years (range 8–86 years); 22 patients were male and 64 patients were female.

Forty-six patients experienced a single episode of SAH prior to their surgery and 40 patients had more than 1 bleed (2–8 hemorrhages) confirmed by computed tomographic (CT) scan and/or lumbar puncture.

The aneurysms arose at the usual sites from the vertebral-basilar or posterior cerebral arteries, with the basilar bifurcation aneurysm being most common. The distribution of aneurysms on the posterior circulation in this early operative group is similar to our overall larger series (Table 1). Fifty-nine of the aneurysms were small (< 12 mm in diameter), 22 of the aneurysms were large (13–24 mm in diameter) and 5 were giant (> 25 mm in diameter).

Table 1. *Location of the Ruptured Aneurysms on the Posterior Circulation* (treated within 7 days of SAH)

• Basilar		66
- Bifurcation	50	
- SCA	14	
- V-B Junction	2	
• Vertebral		15
• Posterior Cerebral		5
Total		**86**

A CT scan was obtained within 2 hours after the most recent SAH in 60 patients. Using Fisher's classification, the amount of blood seen on the CT was graded at (+) in 9 patients, (+ +) in 21 patients, (+ + +) in 30 patients and (+ + + +) in 10 patients[9]. More than half the patients (46) were grade I (Botterell) at the time they were taken to the operating room, 27 were grade I, 10 were grade III and 3 were grade IV.

Surgical Management and Results

It is obvious that early operation to repair a ruptured intracranial aneurysm is a prophylactic procedure to prevent further hemorrhages. There is widespread concern, however, that early operation is technically more difficult because of a tight, angry, red brain and fresh clot obliterating the surgical planes, making access and exposure more difficult. As well, there is concern that the sac is more friable which may increase the likelihood of intraoperative rupture. Moreover, there is a widely held assumption that cerebral ischemia secondary to reactive arterial narrowing (vasospasm) is more common and more troublesome after early surgery. Even so, it has been postulated that early operation may reduce the incidence of post SAH vasospasm and hydrocephalus as a result of the direct removal of clot at operation. The validity of these notions was evaluated in this review. It should be noted that, although the patients included in this analysis were operated on during a period of 15 years, 85% were operated upon in the past 5 years during an era of modern neuroradiologic evaluation, effective contemporary neuroanesthesia techniques and the routine use of microsurgery.

Because of our increasing confidence in early surgery and because patients with these difficult lesions were being referred to our unit soon after the hemorrhage, the opportunity for early operation has, in recent years, become more commonplace.

Basilar Bifurcation Aneurysms

Of the 50 patients in this study with ruptured basilar bifurcation aneurysms, 28 were grade I, 15 were grade II and 7 were grade III. The neck of the aneurysm was directly clipped in 47 patients but, because of giant size and an associated rigid atherosclerotic neck, 3 patients were treated with upper basilar artery occlusion. All aneurysms of the upper basilar artery were approached by the subtemporal route and in only two instances was intracranial pressure considered high, making this exposure limited. Although retraction of the temporal lobe was somewhat difficult in these two cases, both were grade I at the time they went into the operating room and with the use of Mannitol and lumbar spinal fluid drainage, as well as controlled ventilation, adequate exposure was obtained with complete obliteration of the aneurysms and ultimately excellent results.

Major neurologic dysfunction was noted in the immediate postoperative period in 17 of the 50 patients with basilar bifurcation aneurysms. Analyses of the causes of this neurologic deterioration revealed intraoperative rupture with secondary imperfect neck clipping in three patients, temporal lobe swelling from heavy retractor pressure in one, postoperative extradural hematoma in one, rupture of the aneurysm prior to tourniquet closure in one, and probable perforator injury in the remaining patients. Of these 17 patients who deteriorated from events surrounding the operation, 13 recovered completely within several weeks; 3 remain permanently disabled due to thalamic and mid brain infarction secondary to perforator injury. The patient with the postoperative extradural hematoma expired.

In 10 patients, delayed cerebral ischemia secondary to vasospasm occurred between the seventh and thirteenth days following the first SAH; this was documented by cerebral angiography and cerebral blood flow studies. All were treated with hypervolemic-hypertensive therapy and although three of them recovered completely, five patients developed permanent neurologic deficits and two died from cerebral infarction.

Of the 26 patients operated on within 96 hours of their last SAH, and 24 patients operated on between Day 5 and Day 7, excellent results were obtained in 22 (85%) and 18 (75%) respectively. In this small series, therefore, there appears to be no significant difference in outcome between these two groups of patients, whether operated upon at the beginning or at the end of the first week following their SAH. The operative and postoperative complications were equally divided between these two groups of patients.

Basilar Superior Cerebellar Artery Aneurysms

Fourteen patients presented with rupture of aneurysms arising from the basilar artery at the origin of the superior cerebellar artery. Six of these patients were grade I, 5 were grade II, 2 were grade III and 1 was grade IV at the time they were taken to the operating room. All patients underwent direct clipping of the aneurysm. Five patients were operated on within 96 hours of the SAH and 9 patients between the fifth and seventh days.

All of these aneurysms were approached via the subtemporal route, under the right temporal lobe in six patients and under the left temporal lobe in eight patients, according to the side toward which the aneurysm was pointing. On opening the dura, a tight brain was encountered in two patients. One patient, a Botterell grade II at operation, had an excellent outcome. The other (grade IV) died as a result of imperfect clipping of the aneurysm after troublesome intraoperative hemorrhage obscured the field resulting in a hurried and imperfect placement of the clip.

One other patient, grade I, suffered a major intraoperative hemorrhage and imprecise placement of the clip, with partial occlusion of the superior cerebellar artery and one of its proximal perforators, and remains disabled with a cerebellar infarction. Two patients, both grade III prior to operation, developed delayed ischemia secondary to vasospasm; one of these recovered completely and the other remains partially disabled due to an infarct in the opposite parietal lobe. Immediately after surgery, five patients condition worsened; this may have been a result of the intraoperative trauma of temporary ischemia due to prolonged generalized hypotension or the use of a temporary basilar occlusion or perforator manipulation or injury. However, all the patients recovered completely within two weeks after surgery.

Overall, the results were excellent in 11 patients (79%), poor in 2 and fatal in 1.

Vertebral-Basilar Junction Aneurysms

Two patients with vertebral-basilar aneurysms underwent direct clipping of the aneurysm. One patient (grade I) was operated on 5 days after a third SAH and 12 days after his first hemorrhage. The aneurysm was large, complex and required multiple clips to secure the neck. Postoperatively, he was noted to have dysfunction of the sixth, seventh and eighth cranial nerves on the operative side which gradually recovered over several months. He has returned to full activity, but has been left with a significant impairment of hearing in the left ear. In the other patient, grade IV at the time of operation and 1 day after her fourth SAH and 15 days after her first hemorrhage, the aneurysm was secured uneventfully and subarachnoid clot removed. Even so, she developed widespread cerebral ischemia 3 days postoperatively and 18 days after the first SAH; despite vigorous attempts at improving perfusion, she died.

Vertebral Aneurysms

Early operation was performed in 15 patients with ruptured vertebral aneurysms. Eight patients were grade I, 6 were grade II and 1 was grade IV at the time of operation. Fourteen of the 15 patients had aneurysms arising at the vertebral PICA junction and 1 arose from a large perforator distal to the origin of PICA. Fourteen of the patients underwent direct clipping of the neck of the aneurysm and one was treated with proximal vertebral artery occlusion. Seven patients were operated on within 96 hours of the SAH and the remainder between Day 5 and Day 7.

Eight patients developed postoperative neurologic deficits. Four had transient lower cranial nerve dysfunction which recovered without disability over the next several months. Three patients had evidence of cerebral ischemia secondary to vasospasm and all recovered completely. One patient, who had a large vertebral PICA aneurysm and was grade IV at the time of operation following two hemorrhages and marked communicating hydrocephalus, had the aneurysm secured without difficulty and was unchanged immediately postoperatively, but developed an acute subdural hematoma following the insertion of a ventriculostomy to control the hydrocephalus and died. This tragic outcome was the only untoward result in the 15 operative cases.

Posterior Cerebral Artery Aneurysms

Five patients presented with rupture of posterior cerebral artery aneurysms; three aneurysms arose from the P 1 segment and two from the P 2 segment. Three patients were grade I, 1 was grade II and 1 was grade III at the time of operation. The aneurysms were clipped within four days of the most recent SAH in two patients and between Day 5 and Day 7 in three patients. Immediately postoperatively, two of the patients with P 1 aneurysms were noted to have a hemiparesis, both thought to be due to injury to the long perforating branches arising from the P 1 segment. One of these patients recovered completely and the other remained disabled with a posterior thalamic infarction. One patient with a P 2 aneurysm and grade II, had delayed hemiparesis and aphasia on Day 12. She responded to intravascular volume expansion and elevation of blood pressure, recovering completely. Ultimately, four of the five patients had excellent results.

Relationship of Preoperative Timing and Results

The overall results designated by grade at operation are summarized in Table 2. At the time of follow-up, at least one year after their operation, patients were categorized as excellent only if they were neurologically normal on repeat examination. Patients with detectable neurologic abnormalities but who had returned to work and to an unrestricted lifestyle were classified as good. The "poor" categorization describes patients with detectable neurologic dysfunction who were unable to work or required some level of assistance in their daily living. It should be noted that all of the grade IV patients died and 30% of the grade III patients at operation ultimately were significantly disabled or dead. If one considers only the grade I and II patients, however, 10% of the patients were ultimately disabled and less than 3% of them expired. When examining only the grade I and II patients (Table 3) and comparing the results with those operated on within 4 days of their most recent SAH and those operated on between Day 5 and Day 7, it is noted that a good or excellent outcome was obtained in 93% and 81% respectively. By comparison, we have operated on 888 patients with a grade I or II coming to the operating room and operated on at least one week after their last SAH (average Day 12) showed a 90% excellent or good outcome. The operative mortality in grade I or II patients, whether operated on in the first week or delayed, was identical at 3%.

Grades III and IV patients operated on early resulted in only a 54% good or excellent outcome (all of these patients being grade III) and a 31%

Table 2. *Results of Surgery for the Posterior Circulation Aneurysms* (treated within 7 days of SAH)

GRADE*	NO.	EXCELLENT	GOOD	POOR	DEAD
I	46	30	9	5	2
II	27	12	12	3	0
III	10	4	3	2	1
IV	3	0	0	0	3
Total	**86**	**46**	**24**	**10**	**6**
		(81%)			(7%)

*Botterell Grading

Table 3. *Results of Surgery for the Posterior Circulation Aneurysms*

	TIMING	NO.	EXCELLENT/GOOD	DEAD
• Grade I, II	<96 hrs.	30	28 (93%)	0
	Day 5-7	43	35 (81%)	2
• Grade III, IV	<96 hrs.	11	5	4
	Day 5-7	2	2	0

Table 4. *Results of Surgery for the Ruptured Posterior Circulation Aneurysms*

GRADE	TIMING OF SURGERY	NO.	EXCELLENT OR GOOD	DEAD
• I, II	< 7 days	73	86%	3%
	Elective	888	90%	3%
• III, IV	< 7 days	13	54%	31%
	Elective	132	57%	15%

mortality, but as disappointing as these results are, they are not dissimilar to those poor-grade patients undergoing their operation electively in whom only 57% were considered good or excellent at the time of follow-up and 15% had died as a result of their SAH or operation (Table 4). Although all of the grade IV patients subjected to early surgery died, only two of the grade IV patients operated on electively had a good outcome, with five patients being significantly disabled and three dying.

Aneurysm Size and Results

From our experience, it would appear that aneurysm size, up to 24 mm in diameter, has little effect on outcome. Good or excellent results were obtained in 83% of the small aneurysms and 82% of the large aneurysms, with a 7% and 9% mortality respectively in those patients that were operated on early. In those patients undergoing delayed surgery, a 91% good or excellent result was obtained in small aneurysms and 82% in large aneurysms with a mortality of 4% and 3%. Because of the large difference in the numbers in the two groups, these percentage differences are likely not significant. Five patients with giant aneurysms underwent early operation with excellent or good results in 60%, which is similar again to those operated on electively in our unit.

Sixty of the patients operated on early were less than 60 years of age and 18 patients were 60 years of age or older. Slightly better operative results were obtained in the younger patients, with 84% of these having an excellent or good result and a 6% mortality as compared to the older patients, who suffered 11% mortality and only 72% had a good or excellent result. Again, these comparisons suffer from the relatively small number of patients and by the fact that there were a higher proportion of poorer grade patients in the older group.

Forty-six patients were operated on early after a single SAH; 82% of these had a good or excellent outcome and 7% died. Similar results were obtained in the 40 patients who had multiple hemorrhages; 80% had a good or excellent outcome and 8% died.

Communicating hydrocephalus, secondary to SAH, was clinically significant and radiologically confirmed in 11 of the 86 patients undergoing early surgery. Nine of the 11 were patients who had had multiple episodes of SAH and ultimately 4 patients required operative CSF diversion. One developed an acute subdural hematoma following ventriculostomy as noted above.

Intraoperative rupture of the aneurysm occurred in 10 patients. In four instances, the aneurysm arose from the basilar bifurcation, in five the aneurysm was at the basilar superior cerebellar artery origin and one arose from the vertebral artery at PICA. The intraoperative rupture was considered minor in five patients. Two of these with minor hemorrhages had an uneventful postoperative course. Two other patients went on to develop delayed ischemia secondary to vasospasm; one of them recovered completely and the other partially. One patient was more drowsy postoperatively, thought due to heavy retraction and temporary clipping of the basilar artery, but eventually recovered completely. In contrast, five patients sustained a major intraoperative rupture of the aneurysm and all were immediately worse as a result of the operative experience. Only two of these patients ultimately made an excellent recovery. One patient, who clearly suffered from an improperly placed clip and perforator injury, remained disabled. Two patients who were poor-grade going into the operation both unfortunately experienced an imperfect clip application and ultimately died.

Incomplete Obliteration of the Aneurysm

Of the 82 patients who underwent direct clipping of the aneurysm, the neck occlusion was incomplete in 3 patients. Two of these patients harbored large basilar bifurcation aneurysms and one had a large basilar superior cerebellar artery aneurysm. Re-operation and repositioning of the clip was carried out in two patients, with an excellent result in one. The other, however, who was grade IV at the time of the first and second procedure, ultimately died. The third patient, with a large basilar bifurcation aneurysm, also grade IV at the time of the initial procedure, did not improve following the second procedure and was not considered in sufficiently good condition after a delay of two weeks to warrant further surgical exploration.

Proximal vessel occlusion was performed in four patients when it was proved to be impossible to directly clip the neck of the aneurysm. This resulted in complete thrombosis of the aneurysm in three instances with good results. In one patient with a giant basilar bifurcation aneurysm, a tourniquet was used to close the basilar artery the day after it was placed around the vessel and the patient tolerated this well. However, 24 hours later, this patient suffered a major rebleeding despite a completely occluded basilar artery and has remained neurologically disabled. No further attempts to prevent flow into this aneurysm have been undertaken.

Vasospasm

Delayed ischemic neurologic deficit secondary to vasospasm was identified as the cause of neurologic deterioration in 14 patients (16%).

Fourteen of the 86 patients (16%) developed delayed neurologic ischemic deficit as a consequence of reactive arterial narrowing or vasospasm. The onset of these ischemic symptoms ranged between 3 and 14 days following the last SAH. Vasospasm was documented in 11 of the 73 (15%) grade I and II patients and in 3 of the 13 (23%) grade III and IV patients. All patients were started on a regimen of hypervolemia and hypertension as soon as the diagnosis of vasospasm was made and competing other causes of neurologic deterioration either ruled out or remedied. Five patients with confirmed vasospasm responded promptly to treatment with excellent results. Two patients recovered more slowly, but were judged to have excellent or good results at the time of follow-up in one year. Four patients improved as the result of the hypervolemia/hypertensive therapy, but not before the impaired perfusion had produced cerebral infarction and, although they survived, they remained disabled. Three patients died as a direct consequence of the vasospasm.

In 60 patients, a CT scan was performed within 12 hours of their SAH. The amount of blood in the subarachnoid cisterns was graded according to the scale of Fisher. Of 30 patients with a CT grade of (+) and (+ +), only 2 (7%) developed delayed cerebral ischemia secondary to vasospasm and both responded promptly to therapy and demonstrated a complete recovery of neurologic function before discharge from hospital. Of the 30 patients with a CT grading of (+ + +) and (+ + + +), 9 developed delayed neurologic deficit (30%). Despite vigorous attempts to expand their blood volume and elevation of blood pressure, 5 remained permanently disabled or died (Table 5). Of the patients with CT grade (+) or (+ +), 93% had a good outcome and no mortality, but of the patients with a more massive bleed, as evidenced by a large amount of clot in the basal cisterns, only 77% had a good outcome, and there was a 10% mortality (Table 6).

This incidence and outcome of the complication of vasospasm is identical to our experience with the management of SAH and vasospasm associated with anterior circulation aneurysms in our unit in the past five years. Moreover, the incidence of vasospasm does not appear to have been increased as a result of early operation. Although early operation provides the immediate security of an obliterated aneurysm when one anticipates the possibility of delayed ischemia secondary to vasospasm, this small series does not suggest that early operation and vigorous therapy to improve perfusion has resulted in a significantly better outcome.

Table 5. *Delayed Neurological Deficit Following Rupture of the Posterior Circulation Aneurysms*

	NEUROLOGICAL DEFICIT		
CT GRADE*	TRANSIENT	PERMANENT OR DEAD	TOTAL
+, + + (n = 30)	2	0	2 (7%)
+ + +, + + + + (n = 30)	4	5	9 (30%)
Total	6	5	11 (18%)

\+ : No subarachnoid blood visualized
\+ + : Diffuse or thin sheet
\+ + + : clot or thick layer
\+ + + + : Diffuse or none with intracerebral or intraventricular clots

(Fisher)

Table 6. *Results of Surgery for the Posterior Circulation Aneuryms* (within 7 days of SAH) *by CT Grade*

CT GRADE	NO.	EXCELLENT	GOOD	POOR	DEAD
+, ++	30	17	11	2	0
		93% (excellent + good)			
+++, ++++	30	16	7	4	3
		77% (excellent + good)			10%

Discussion

Acute surgery for posterior circulation aneurysms was attempted in the earliest experience of the senior author. Two of the first four patients died, one with a cardiac arrest, following poorly controlled hypotension and hypothermia[3]. Since then and after operating on more than 1400 patients in our unit with aneurysms arising from the vertebral and basilar arteries, the pre- and postoperative care, techniques of neuroanesthesia and operation, and the recognition and management of complications of SAH have become established and quite routine[4–7, 20, 21]. Overall, 85% of these patients have achieved excellent or good results and the mortality for the whole group is 5%. Even so, we continue to be concerned by a combined morbidity and mortality of 15% and are continually haunted by the deterioration and death of patients because of recurrent hemorrhage before the aneurysm could be secured. Most surgeons are unfamiliar with aneurysms hidden in front of the brain stem. As a result, there has been a general trend, supported by our previous publications, to delay surgery for those aneurysms arising from the posterior circulation at a time when there has been a trend toward early surgery for ruptured aneurysms arising from the anterior circulation[1, 7, 11, 13, 16, 22, 24, 26]. In our unit (Table 7), we have compared the results of early operation (within 96 hours) in patients with aneurysms arising from the anterior circulation and those with aneurysms arising from the posterior circulation. The results are equally good in those patients graded Botterell I or II at the time of surgery and equally poor in those graded III or IV. Thus, in our hands, the condition of the patient going into the operating room would appear to be the most powerful determinent of outcome.

It has been suggested that vasospasm is less common following rupture of aneurysms of the posterior circulation. It has also been suggested that

Table 7. *Results of Early Surgery for Ruptured Aneurysms*

GRADE (BOTTERELL)	LOCATION OF ANEURYSMS	NO.	EXCELLENT OR GOOD	DEAD
I, II	Anterior Circulation	41	93%	2%
	Posterior Circulation	30	93%	0%
III, IV	Anterior Circulation	8	50%	25%
	Posterior Circulation	11	45%	31%

early operation may increase the incidence of vasospasm by some unknown effect of the surgical manipulation or anesthetic carried out at a time when the vessel's reactivity is most vulnerable. Alternatively, there has been hope that the incidence of vasospasm may be reduced by virtue of the evacuation of clot and bloody CSF afforded at the time of the surgical intervention. Our experience would not support any of these notions. The incidence of vasospasm following SAH in anterior and posterior circulation aneurysms is 15% and 16% respectively in those cases undergoing early surgery. Overall, the incidence of vasospasm in our larger group of patients, where surgery has been delayed, is 15%. Whether early or delayed, 50% of the patients with ischemic dysfunction secondary to vasospasm recover with excellent or good results with hypervolemic/hypertensive therapy. The remaining half (7% of the total with vasospasm) are left disabled or dead as a result of their cerebral infarction. The most powerful predictor of vasospasm has been the CT documentation of subarachnoid clot. Although not perfect, the predictive value of CT imaging of clot in the subarachnoid cisterns is high, and more importantly, those patients with minimal or no clot in the cisterns following the bleed that do go on to develop vasospasm and ischemia are most likely to respond dramatically to the elevation of their intravascular volumes and blood pressure and improvement of the cerebral perfusion pressure[14, 19].

Isolation of the thin sac of an aneurysm by surgical obliteration of the neck is the ultimate goal of the surgical treatment of intracranial aneurysms to prevent rebleeding. If the sac is large or of giant size, the technical complexity and relative inaccessibility of these aneurysms is markedly increased and usually will make neck clipping difficult if not impossible and certainly hazardous for the patient. In the past, we have suggested that proximal or hunterian ligation in these circumstances is an effective and usually safe alternate method of treatment in those patients coming electively to surgery[7,18]. Proximal vessel ligation was attempted in only four patients in the acute stage and only two benefited from the therapy. One fatality and one poor outcome occurred in patients grade I and II respectively harboring giant aneurysms. These disappointing results in a few patients have made us cautious about attempting proximal vessel ligation in the immediate week following acute SAH.

Several reports have suggested a high mortality (48—83%) due to rebleeding from aneurysms arising from the posterior circulation[8,10,17,25]. In more than 1400 aneurysms of the posterior circulation operated upon by the senior authors, 1126 had ruptured and 86% had good or excellent result with an operative mortality of 5%. In those patients operated on within 7 days of the SAH, 81% had functional results and 86% of those patients grade I or II were considered excellent or good at follow-up. Whether the aneurysm was judged to be small or large had no effect on outcome. Only three of the five giant aneurysms operated on early had a good result.

In our hands, early surgery for aneurysms of the posterior circulation yields results similar to early surgery for aneurysms of the anterior circulation and not dissimilar to the results achieved in delayed operation of aneurysms of the posterior circulation in our unit. It is clear that early surgery, with the aneurysm appropriately clipped, will reduce dramatically the risk of rebleeding. Moreover, our review would suggest that the incidence of vasospasm is not more common or less amenable to therapy or more severe in those patients who have been operated on early. Indeed, it has seemed logical to us that the only therapy appearing to benefit those patients with impaired cerebral perfusion secondary to vasospasm—namely volume expansion and elevation of blood pressure—is more safely administered with the aneurysm surgically obliterated. Although early surgery more frequently produces the initial impression that the brain is more swollen and tight and the exposure more confining, the use of lumbar spinal fluid drainage, controlled ventilation and osmotic diuretics has ultimately made the exposure possible. As well, dissection of the aneurysm itself with removal of surrounding fresh clot, identification of the neck and separation of perforators, although somewhat more tedious in the acute case, is neither significantly difficult nor more hazardous. Intraoperative aneurysmal rupture was again not apparently different in those cases operated on early or late.

This review, as in the International Collaborative Study of the Timing of Aneurysm Surgery, does not appear to show a significant difference in outcome between those patients operated on early or late when presenting with ruptured posterior circulation aneurysm[15]. These conclusions are marred by the fact that we are comparing possibly two different groups of patients and the numbers of those operated on acutely in the London, Ontario series are distinctly small.

We have modified our approach over the years to operate early in those patients who are good-grade, whose aneurysms do not present a particular technical difficulty because of size, configuration or location, and occasionally in those patients whose lives appear to be in jeopardy because of recurrent hemorrhage. An overriding consideration, however, is that we would not recommend and rarely attempt one of these procedures in the middle of the night or on weekends or holidays when it may prove to be impossible to have a fresh, rested and experienced team of nurses, anesthetists and surgeons to conduct these frequently difficult procedures. We are certain that the availability of an experienced, alert, surgical team operating on a good-grade patient is the most important determinant for a good outcome.

References

1. Auer LM (1983) Acute surgery of cerebral aneurysms and prevention of symptomatic vasospasm. Acta Neurochir (Wien) 69: 273–281
2. Dinsdale H, Logue V (1959) Ruptured posterior fossa aneurysms and their surgical treatment. J Neurol Neurosurg Psychiat 22: 202–207
3. Drake CG (1961) Bleeding aneurysms of the basilar artery. Direct surgical management in four cases. J Neurosurg 18: 230–238
4. Drake CG (1965) Surgical treatment of ruptured aneurysms of the basilar artery. Experience with 14 cases. J Neurosurg 23: 457–473
5. Drake CG (1968) Further experience with surgical treatment of aneurysms of the basilar artery. J Neurosurg 29: 372–392
6. Drake CG (1968) The surgical treatment of aneurysms of the basilar artery. J Neurosurg 29: 436–446
7. Drake CG (1979) The treatment of aneurysms of the posterior circulation. Clin Neurosurg 26: 96–144
8. Duvoisin RC, Yahr MD (1965) Posterior fossa aneurysms. Neurology (Minneap) 15: 231–241
9. Fisher CM, Kistler JP, Davis JM (1980) Relation of cerebral vasospasm to subarachnoid hemorrhage visualized by computerized tomographic scanning. Neurosurg 6: 1–9
10. Hook O, Norlen G, Guzman J (1963) Saccular aneurysms of the vertebrobasilar arterial system. A report of 28 cases. Acta Neurol Scand 39: 271–304
11. Hori S, Suzuki J (1979) Early intracranial operations for ruptured aneurysms. Acta Neurochir (Wien) 46: 93–104
12. Jamieson KG (1964) Aneurysms of the vertebrobasilar system. Surgical intervention in 19 cases. J Neurosurg 21: 781–797
13. Kassell NF, Boarini DJ, Adams HP *et al* (1981) Overall management of ruptured aneurysm: comparison of early and late operation. Neurosurg 9: 120–128
14. Kassell NF, Peerless SJ, Durward QJ *et al* (1982) Treatment of ischemic deficits from vasospasm with intravascular volume expansion and induced arterial hypertension. Neurosurg 11: 337–343

15. Kassell NF, Torner JC, Jane JA (1985) International Cooperative Study on Timing of Aneurysm Surgery—Preliminary Report. Early versus Late Planned Surgery. Presented at the American Association of Neurological Surgeons Annual Meeting, Atlanta, Georgia
16. Ljunggren B, Brandt L, Sundburg G *et al* (1982) Early management of aneurysmal subarachnoid hemorrhage. Neurosurg 11: 337–343
17. Locksley HB (1969) Natural History of Subarachnoid Hemorrhage. In: Sal AL, Perret GE, Locksley HB *et al* (eds) Intracranial aneurysms and subarachnoid hemorrhage. Lippincott, Philadelphia, pp 37–108
18. Mount LA, Taveras JM (1962) Ligation of basilar artery in treatment of an aneurysm at the basilar artery bifurcation. J Neurosurg 19: 167–170
19. Peerless SJ (1979) Pre- and postoperative management of cerebral aneurysms. Clin Neurosurg 26: 209–231
20. Peerless SJ, Drake CG (1985) Posterior circulation aneurysms. In: Wilkins RH, Rengachary SS (eds) Neurosurgery. McGraw-Hill, New York, pp 1422–1437
21. Peerless SJ, Drake CG (in press) Mangement of aneurysms of posterior circulation. In: Youmans JR (ed) Neurological surgery, 3rd edn. Saunders, Philadelphia
22. Sano K, Saito I (1978) Timing and indication of surgery for ruptured intracranial aneurysms with regard to cerebral vasospasm. Acta Neurochir (Wien) 41: 49–60
23. Sugita K, Kobayashi S, Shintani A *et al* (1979) Microneurosurgery for aneurysms of the basilar artery. J Neurosurg 51: 615–620
24. Sundt TM, Kobayashi S, Fode NC *et al* (1982) Results and complications of surgical management of 809 intracranial aneurysms in 722 cases. J Neurosurg 56: 753–765
25. Troupp H (1971) The natural history of aneurysms of the basilar bifurcation. Acta Neurol Scand 47: 350–356
26. Wilson CB, Hoi Sang V (1976) Surgical treatment for aneurysms of the upper basilar artery. J Neurosurg 44: 537–543
27. Yaşargil MG, Antic J, Laciga R *et al* (1976) Microsurgical pterional approach to aneurysms of the basilar bifurcation. Surg Neurol 6: 83–91

Neuro-Anaesthesia: the Present Position

DOREEN JEWKES

Department of Neurological Surgery, Institute of Neurology, The National Hospital, London (U.K.)

With 3 Figures

Contents

Anaesthesia and analgesia for neurosurgical procedures is subject to certain physiological and pharmacological principles which, if ignored, will lead to poor operating conditions for the surgeon or an unnecessarily damaged patient. The fundamental principle underlying all anaesthesia must never be obscured. The anaesthetic state of the patient must be associated with the delivery of adequate oxygen to all cerebral tissue. This requires particular attention to the cerebral circulation and how it is affected by pathological processes, surgical intervention and the anaesthetic techniques.

Working according to these principles demands the continuous attention of the anaesthetist to the patient directly and to the electronic displays with their elegant representation of cardiovascular and respiratory function. Modern displays provide the anaesthetist with a continuous record of pulse rate, blood pressure, electrocardiogram and arterial wave forms, C.V.P. if desired and end-tidal CO_2. This list seems to grow yearly!

Physiological and Pathophysiological Considerations

Cerebral blood flow affects not only oxygen delivery and removal of metabolic products, but also cerebral volume and there by intracranial pressure. There are some situations in which an increase in flow can lead to ischaemia and, therefore, sub-oxygenation of brain cells. An understanding of cerebral flow in physiological and pathophysiological conditions is absolutely necessary to the neuro-anaesthetist.

Normal cerebral blood flow values for man are 18–20 ml/100 g^{-1}/min^{-1} for white matter and 78–80 ml/100 g^{-1}/min^{-1} for grey matter. This is relatively constant despite changes in arterial blood pressure within the normal range. Autoregulation maintains flow constant when the mean arterial pressure varies between 50–150 mg/Hg. The mechanism of autoregulation is not precisely known, but with changes in blood pressure there are reciprocal changes in local cerebro-vascular resistance. There is a rapid decrease in cerebral blood flow if the arterial pressure falls below this range. Above 150 mm/Hg mean pressure cerebral blood flow increases (autoregulatory breakthrough) and oedema fluid may form (Skinhoj and Strandgaard 1973).

In the severely hypertensive patient, the lower threshold of autoregulation is higher than 50 mm/Hg, which has important practical consequences (Strandgaard, Olesen, and Skinhoj 1973). The pressure which drives blood flow to the brain is the mean arterial pressure minus the sum of the intracranial pressure (I.C.P.) and is normally about 80 mm/Hg. In the presence of raised I.C.P. cerebral perfusion pressure is maintained by increasing arterial blood pressure, either spontaneously or as a deliberate therapeutic manoeuvre. This compensatory mechanism fails when the I.C.P. is in the region of 30–40 mm/Hg. With neuro-anaesthesia, raised

I.C.P. makes a patient vulnerable from induced hypotension to any degree. However, this is usually only an anxiety until such a time as the surgeon has raised the craniotomy flap.

The practical aim of the neuro-anaesthetist is to maintain the arterial pressure within the range of normal cerebral autoregulation and to counteract any influences, physiological or pharmacological, which might tend to increase or decrease cerebral perfusion. Cerebral blood flow increases when cerebral blood vessels dilate. Vasodilation is caused by hypercapnia, anoxia and trauma, also by most anaesthetic agents. Hypocapnia conversely causes vasoconstriction of healthy vessels thereby tending to reduce intracranial pressure.

On balance the majority of influences operating in the anaesthetised patients are working in the direction of increasing cerebral blood flow so that the patient is best maintained in a state of hypocapnia. This will produce a baseline cerebral vasoconstriction. The arterial CO_2 of neurosurgical patients is therefore maintained between 25–30 mm/Hg (4 kPa) during surgery, and in all patients with actual or potential brain damage from haemorrhage, tumour or trauma. Blood flow may be increased to damaged or compressed brain by the constriction of healthy normal vessels elsewhere in the brain. In pathological states vasomotor tone is changed and autoregulation and the normal response to CO_2 is lost. Hyperventilation causes normal vessels to constrict and this increases the flow to abnormal areas in which the CO_2 response is absent or impaired producing the so called inverse steal phenomenon as described by Lassen and Christensen (Lassen and Christensen 1976).

Volatile anaesthetic agents are all vasodilators, so cerebral blood flow and therefore brain volume will increase when they are used in an anaesthetic technique. However, generalised vasodilation means that the arterial blood pressure will fall, limiting the increase in cerebral blood flow in areas with impaired autoregulation (Fig. 1, p. 134).

Volatile anaesthetic agents should be introduced into the anaesthetic technique only after hypocapnia is established at 25–30 mm/Hg (4 kPa). Their vasodilating effect on cerebral vessels is therefore greatly modified. All anaesthetic agents and techniques are chosen with these physiological principles in mind, considering also the patients age, pathology and presence of pre-existing disease and the position of the patient as required for neuro-surgery.

Hypoxia can be catastrophic to the neurosurgical patient. If PaO_2 falls to 50 mm/Hg (6.5 kPa) vasodilation occurs and cerebral blood flow rises, increasing intracranial volume and pressure (Cohen *et al.* 1967).

At the end of surgery haemostasis is also benefited by the same physiological mechanism. As the $PaCO_2$ rises on cutting ventilation or reversing neuromuscular blockade to enable the patient to breathe sponta-

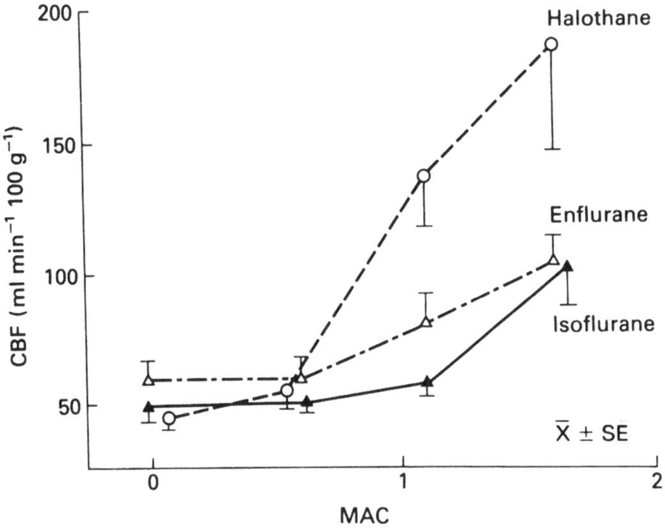

Fig. 1. CBF measured in normocapnic volunteers awake and anaesthetized with nitrous oxide and different volatile agents. Systemic pressure was maintained at normal levels by an infusion of phenylephrine and normocapnia by ventilation. From Eger (1981)

neously, vasodilation occurs in normal brain tissues. Intracerebral steal occurs, blood flowing preferentially to normal areas rather than operated areas assisting surgical haemostasis; this effect is enhanced by vasodilation produced by inhalational agents.

Practial Aspects/Considerations

Premedication: Neurosurgical patients are rarely heavily sedated. The anaesthetist's aim is to gain the confidence of the patient, giving reassurance by a pre-operative visit and only prescribing sedatives and tranquillisers when absolutely necessary. Premedication is therefore oral, and restricted to benezodiazepines or barbiturates, phenobarbitone remaining a useful medication. Narcotics are never used, so pupillary signs in the post-operative period are therefore reliable and there is no danger of respiratory depression.

Medical treatment for pre-existing conditions, *e.g.* diabetes mellitus or hypertension is continued. Physiotherapy is given to all patients with pre-existing chest problems. Psychiatric patients are also anaesthetised without withdrawal of their anti-depressant drugs. Most operations are urgent and cannot be delayed for the two weeks needed to eliminate monoamine-oxidase inhibitors from the body.

The anaesthetist is concerned that patients who are receiving anti-convulsants either because they have already had a seizure, or because they are to have a supratentorial craniotomy, have achieved a serum level in the therapeutic range. Patients receive dexamethasone as part of their routine preparation for operation. It is wise to prescribe an H_2 antagonist when giving a relatively large dose of steroids. Cimetidine enhances the sedation produced by benzodiazipines, but ranitidine does not seem to have this undesirable property.

Induction

The induction agent selected should be pain free on injection. Several new agents have been developed in the past few years but thiopentone 5–10 mg/kg by intravenous injection, is usually the induction agent of choice. It diffuses into the brain and produces its effect in thirty seconds. It is depressant to the myocardium so cardiac output falls, returning to normal in a few minutes, with depression of blood pressure most marked in hypertensives. In the blood thiopentone is bound to plasma proteins. Increased binding occurs as the plasma pH is increased by hyperventilation, the effects of thiopentone are increased by lowering $PaCO_2$. Its depressant effect on ventilation is unimportant, as all patients are intubated and ventilated. There are however, contra-indications to the use of thiopentone, more often in neurological practice than in any other situation. It is contra-indicated in most myopathies, porphyria, severe cardiac disease and hypertension. In these patients the new induction agents have a definite place.

Of the newer induction agents only three at present seem to be of interest, etomidate and propofol(2-6-di-isopropylplenol) and midazolam.

Etomidate

Etomidate 2 mg/kg has advantages over thiopentone. It has great cardiac stability and does not cause histamine release, but moderate to severe involuntary muscle movements occur in a high proportion of patients. These are less troublesome if 50–100 mg of thiopentone are given before the induction dose of etomidate. The disadvantages of an increased incidence of cough and hiccup do not occur in ventilated patients, but vomiting in the post-operative period occurs in 30% of patients. Etomidate is therefore mainly used for patients with cardiac disablement or severe hypertension presenting for neurosurgery.

Propofol

Propofol is a non-barbiturate, rapidly acting induction agent, producing a greater fall in arterial blood pressure than the equipotent dose of thiopentone due to a decrease in systemic vascular resistance. Cardiac output falls slightly, a compensatory tachycardia does not occur. The drug can be used as a continuous infusion, decay is exponential. In unpremedicated patients the induction dose is 2.5 mg/Kg. It promises to be particularly useful given by continuous infusion in the intensive care unit.

Midazolam

Midazolam, a water-soluble benzodiazepine, is non-irritant by injection in doses of 0.07 mg/Kg. Induction is swift and without accompanying movements, and with no significant changes in blood pressure, and only transient tachycardia. The drug produces marked anterograde amnesia of 20–40 minutes, an effect useful if the patient is to have several visits to the operating theatre or pain therapist.

Methahexitone

Methahexitone is still used by some anaesthetists, though its quick recovery period seems unsuited to most neurosurgical procedures. It produces intense pain on intravenous injection, and does nothing to obtund the pressure reflex response to laryngoscopy; it exposes the patient to the ever present danger of awareness during intubation.

Relaxants Old and New

Suxamethonium

A short acting depolarising relaxant retains its place as drug of choice for intubating neurosurgical patients. Its action is seen as fasciculation of muscle fibres as depolarisation occurs, and relaxation occurs within 30 seconds. It is ideal for the swift sequence of laryngoscopy, spray using 2 mls of 4% lignocaine, intubation and immediate hyperventilation. It is definitely the safest drug to use for emergency work. Repeated dosage produces vagal stimulation and bradycardia. Atropine will prevent this and must be used when a difficult intubation is encountered.

Tubocurarine Chloride

The oldest relaxant in use still retains its popularity with neuroanaesthetists. It was first described as arrow poison in 1516 by Asconio Sforza in his De Orbe Novo, based on letters for travellers in what is now Brazil. In 1942 Griffith and Johnson in Canada first used it to produce relaxation in anaesthesia. Pharmacologically tubocurarine has several effects which are of interest to the anaesthetist. It is not cardiostable as it selectively blocks autonomic ganglia, sympathetic block being greater than parasympathetic block. Thus, it produces a dose dependant fall in blood pressure of about 20% of resting level. This action is a great use in anaesthesia, particularly as it is potentiated by halothane. The combination of these drugs produces basal hypotension, which can be enhanced by the action of more specific hypotensive agents. Tubocurarine is rarely used in the elderly, hypertensives, neonates and patients with renal failure, the more recently developed relaxants to be described having many advantages.

Atracurium Dibesylate

Atracurium now appears to be the relaxant of choice in the very old, when cardio-stability is important, or in patients with impaired renal or hepatic function (Payne and Hughes 1981). Atracurium has a place in intubation and consequential relaxation. It is the drug of choice when suxamethonium is contraindicated, *e.g.*, muscle diseases or a family history suggestive of malignant hyperpyrexia. It is a competitive blocker of twenty to forty minutes duration. Excellent relaxation for intubation is obtained after 90–120 seconds at a dosage of 0.6 mg/Kg. Used as the sole relaxant it is best administered as a continuous infusion at the rate of 0.5 mg/Kg/hr. Atracurium has advantage of being cardiostable, but is capable of causing bradycardia (Carter 1983) and histamine release has been reported. A precipitate is often formed when an indwelling venous access is not flushed between drugs, an anaphylactic action can occur with or without inclusion of the patients plasma proteins. This response can be marked when atracurium is injected after thiopentone, the precipitate will initiate acute and often life threatening bronchospasm when swept to lung capillaries (Watkins 1986). Its action is potentiated by halothane and hypothermia, so relaxation must be monitored by a nerve stimulator, an anologue display of the strength of 4th-1st contraction in a train of four allows continuous monitoring of the recovery of neuromuscular blockade. Atracurium undergoes spontaneous breakdown into inactive constituents at body temperature and physiological pH (Hoffman elimination). It is a competitive blocker and can therefore be easily reversed by neostigmine.

Vecuronium

Is a drug to be considered when a stable cardiovascular system is all important, and hypotension can be produced by an inhalational agent alone. It is the preferred relaxant for geriatric patients and asthmatics, as there is no record of histamine release, and in particular for those who are already using an "on demand" type pacemaker, as it has no ganglion blocking action and the dose required to produce vagal blockade is very much greater than that needed for neuromuscular blockade.

Pancuronium

Pancuronium still has a place as a useful long-acting relaxant, particularly when the patient is in poor physical condition, and in particular for spinal surgery. Its sympathomimetic action is then welcomed as an aid to maintaining blood pressure. This feature makes it the relaxant most used and suitable for the X-ray department, when studies are prolonged.

Inhalational Agents

In 1883 Sir Victor Horsley investigated the use of chloroform, ether and morphine sulphate for intracranial surgery (Fig. 2).

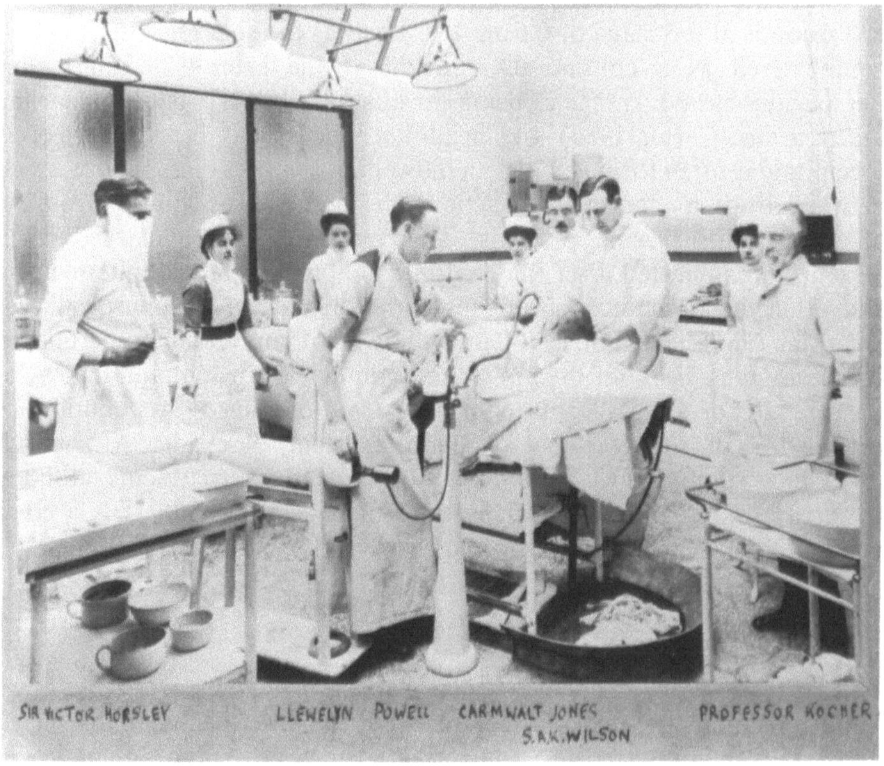

Fig. 2.

He discarded morphine noting that although there was less haemorrhage, its depressant effect on respiration was a distinct disadvantage. He concluded that ether increased arterial pressure and caused excessive bleeding. Chloroform was therefore in the end the favoured anaesthetic agent in the United Kingdom. In the United States of America, Harvey Cushing decided local anaesthesia was greatly to be preferred.

Trichloroethylene

Trichloroethylene (Trilene) was used with nitrous oxide by D. E. Jackson in the U.S.A. in 1934, and has been in general use in this country since 1941. In addition to its anaesthetic effect it has a powerful analgesic action, producing some analgesia into the post-operative period. Tricholoroethylene has a minimal depressant effect on cardiac muscle and blood pressure does not fall, though the pulse may slow slightly. Cardiac muscle becomes more irritable and dysrrhythmias are commonly seen when spontaneous respiration forms part of the anaesthetic technique. These are due to hypercapnia and disappear on lowering the $PaCO_2$. It is usual to give neuromuscular blocking drugs and ventilate all patients anaesthetised with Tricholoroethylene as the volatile agent, controlling also the marked tachypnoea seen with this agent. This tachypnoea is probably of central origin, and is commonly seen while administering inhalational agents (Vickers, Schnieden, and Wood-Smith 1984).

Adrenaline infiltration must be used with caution in any patient anaesthetised with tricholoroethylene and a dose of 20 ml of 1 : 200,000 should not be exceeded. Trichlorethylene, like other inhalational agents increases the irritability of the ventricular conducting conducting system, rendering the beta-adreno-receptors more sensitive to stimulation by catecholamines. If tachycardia occurs, cardioselective beta-blocking agents such as practolol or propranolol should be given to reduce the heart-rate. Trilene has retained a small place in neuro-anaesthesia because of its useful properties of marked analgesia, maintaining a stable cardiovascular system in the sitting position, and is free of explosive risks. It is in addition, very cheap! It is still to be recommended in the radioloy department, for the old and debilitated, and in the sitting position, as its use is consistently trouble free.

In our unit we still use trichloroethylene in the anaesthetic room, following an intravenous induction and intubation, and hyperventilation. The patient then has a stable cardiovascular system while arterial, intravenous and central lines are inserted, and monitoring electrodes for evoked potentials put in place. 0.5% trichloroethylene in no way changes electrophysiological studies, and recordings from the monitor are therefore truly basal.

Halothane

Knowledge of the anaesthetic properties of chloroform, the simplest halogenated hydrocarbon, led to the production of other halogenated hydrocarbons with anaesthetic properties. Halothane was the first such agent to be useful, introduced in the late 1950s. Inspite of some unpromising attributes it is still much used in neurosurgical practice, its good points far outweighing the disadvantages. Its disadvantages are its property of increasing cerebral blood flow, and the known fact that a very small number of patients will suffer liver damage if they are exposed to halothane more than once within twelve weeks. Inspite of these disadvantages, halothane is widely used in neuroanaesthetics at some time during the anaesthetic sequence. The pharmacology and clinical use of halothane has been extensively discussed and has no place in this text (Sadove and Wallace 1962).

Halothane is less soluble in the phospholipids of brain cells than in the neutral fats of adipose tissue. It has a low blood/gas coefficient, so patients recover consciousness fairly rapidly. It is well known that halothane increases cerebral blood flow (C.B.F.) (MacDowall 1967).

The normal cerebral autoregulating mechanism is gradually lost as concentration increases, C.B.F. becoming pressure dependent. Autoregulation is present at 0.5% inhaled concentration but at 1% and 2% it is completely absent (Miletich et al. 1976). At 1% halothane at normocapnia, C.B.F. is increased by 27% and cerebral metabolic consumption of O_2 decreased by 26% compared with awake subjects (Christensen et al. 1967). Recent work has shown that over some hours cerebral blood vessels adapt to halothane by raising cerebrovascular resistance causing a fall in C.B.F. to control values. Autoregulation may remain impaired up to four hours after withdrawing the halothane (Albrecht, Miletich, and Madelar 1983).

Halothane is the inhalational agent of choice for many procedures, being a most useful anaesthetic drug when hypotension is demanded. Hyperventilation must be established for at least fifteen minutes before halothane is introduced to the patient. The $paCO_2$ of such a patient will be 4 KPa (28–30 mm/Kg) or lower. On introducing halothane, general vasodilation and cardiac depression takes place. There is reduction in cardiac output, stroke volume and myocardial contractability, and blood pressure falls, so any tendency to a rise in intracranial pressure is thus prevented (Adams, Gronert, Sundt jr., and Michenfelder et al. 1972). This tendency is more marked in association with an intracranial tumour or haematoma, and can be disastrous in the presence of hypoventilation (Jennett, Barker, Fitch, and McDowall 1969).

Halothane depresses the sympathetic system more than the parasympathetic system. It has a slight beta-adrenergic blocking effect. Patients receiving Ca^{++} antagonists and beta blockers may suffer hypotension if halothane is used for anaesthesia. Ca^{++} antagonists interfere with glycogen metabolism and interact with anaesthetic drugs especially volatile agents. Halothane, Enflurane and Isofluane are non-specific calcium antagonists, partly accounting for myodepressant activity. Combination of Ca^{++} antagonist and volatile agent are therefore additive in their effects on systemic vascular resistance and blood pressure. Heart block may occur due to effects on the A-V node and potentiation of neuromuscular blocking drugs due to reduced release of acetylcholine.

Halothane has no effect on choroid plexus glucose metabolism, but decreases C.S.F. production (Vf) by 30%. This effect does not change with time (Artru 1983).

Halothane Hepatitis

It is now accepted that a mild hepatotoxicity occurs in about 20% of patients who receive halothane anaesthesia; there is no immunological or hypersensitivity reaction but serum transaminase enzymes are raised. However, the severe type of liver damage may be immunologically mediated. It is characterized by massive liver cell necrosis and clinical features of a hypersensitivity reaction, *i.e.*, fever, rash and eosinophilia. Its incidence is assessed at between 1 in 10,000 and 1 in 35,000 single anaesthetics, but 1 in 3,500 of multiple exposures to halothane.

In all clinical series there has been a history of drug allergy. Eosinophilia has been found in 20% of cases, serum auto-antibodies *e.g.*, liver-kidney microsomal antibody is associated with severe liver damage. This antibody reacts with halothane altered hepatocyte membrane components, and is demonstrated by immunofluorescence. In the presence of this antibody, hepatocytes from halothane pretreated rabbits are susceptible to cytotoxicity by normal lymphocytes (Williams 1986, Neuberger et al. 1983).

Enflurane

Enflurane has been used in clinical practice since 1970. It is in many circumstances a most useful anaesthetic agent, recovery being more rapid than from halothane. The myocardium is less sensitive to catecholamines during enflurane anaesthesia, and cardiac dysrhythmias are rare. It is therefore an agent to consider when surgery is planned on patients known to have catecholamine secreting tumours, *e.g.* active glomus tumours of the neck and posterior fossa. It is metabolised to free inorganic fluorine—maximal levels of 25 mmols/l have been found. Cousins and Mazze (1973) showed that levels of twice this value could cause nephrotoxicity. There is no recorded case of liver failure due to enflurane. However, there are several problems with the use of enflurane in neuroanaesthesia. The E.E.G. during enflurane anaesthesia shows episodes of paroxysmal activity with periods of burst suppression. These E.E.G. abnormalities are dependent on concentration and are more obvious in the presence of hypocarbia. Neigh, Garman and Harp (1971), and may persist for up to 30 days after its administration (Burchiel et al. 1977).

Enflurane is therefore used sparingly for neuroanaesthesia, as neuroanaesthetists are wary of this ability to produce epileptiform like activity on the electrocephalogram, particularly in the presence of the obligatory hypocarbia.

Enflurane increases the rate of C.S.F. production and I.C.P. markedly for several hours (Artru et al. 1982). This effect may be the result of increased choroid plexus metabolism as the metabolic rate for glucose is significantly increased. It should be avoided in patients with raised intracranial pressure, if more than 2% enflurane (Ethrane) is necessary (Moss et al. 1983, Meyers and Shapiro 1978). The alternative agents halothane and isoflurane are preferable.

Isoflurane

Isoflurane, an isomer of enflurane, has been in general use since 1981. Recovery from anaesthesia is extremely rapid as it is less soluble in blood than enflurane and minimal biodegradation takes place, only about 0.17% is metabolised (Holaday 1975). Isoflurane has so many theoretical and indeed practical advantages, that it has almost displaced halothane in neuroanaesthesia. Even conservative neurosurgeons appreciate the rapid awakening and opportunity for early neurological assessment.

To compare the drugs a definition of the concept of minimum alveolar concentration or M.A.C. must be given. One M.A.C. is the alveolar concentration of inhalational agent needed to prevent movement in 50% of unparalysed subjects when a skin incision is made. M.A.C. value can thus be compared, and similarly the effect of equipotent concentrations of various inhalation agents on cerebral blood flow (Fig. 3).

Isoflurane at one M.A.C. in normotensive, normocapnic volunteers does not increase cerebral blood flow while one M.A.C. of halothane and enflurane does (Murphy, Kennel and Johnstone 1974). At 1.6 M.A.C. the C.B.F. is doubled but with halothane it is quadrupled. The intracranial pressure rise produced at these concentrations depends on the compliance of the brain, I.C.P. rising in less compliant brains. This rise can be prevented or lessened by hypocapnia, which in neuroanaesthesia is always produced before these agents are introduced. At one M.A.C. isoflurane probably produces less impairment of the autoregulatory response than one M.A.C. halothane, at least this has been found in cats by Drummond, Todd and

Shapiro (1983). This property of isoflurane makes it the preferred agent for patients with cerebrovascular disease or hypertension in whom the blood pressure may be very labile, giving some protection if changes in blood pressure occur from surgical stimulation or haemorrhage.

Isoflurane causes no change in the rate of C.S.F. production (Vf) (Artru 1984) maintaining stability of intracranial dynamics. It has little depressant

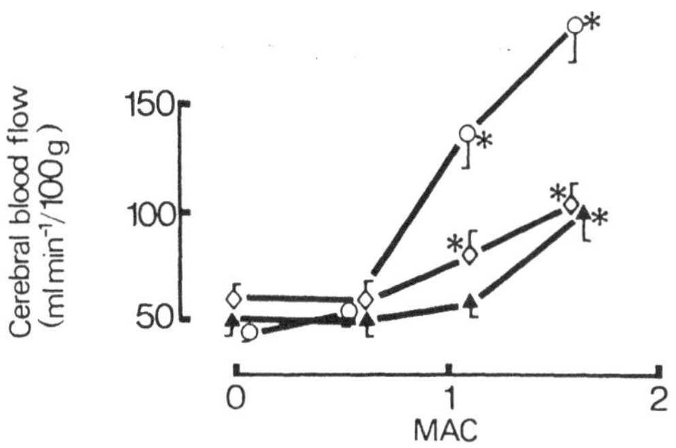

Fig. 3. Cerebral blood flow was measured in volunteers at various values of M. A. C. for three anaesthetic agents. The volunteers were paralysed with tubocurarine and their $paCO_2$ and systemic arterial pressure keptz at normal values. Flow increased at light levels of enflurane (◇) and halothane (○) anaesthesia, but did not increase at the same levels of isoflurane (▲). All three agents increased flow at 1.6 M. A. C. (Data from Murphy, Kennel and Johnstone (1974). Reprinted by permission of Ohio Medical Anesthetics)

effect on baroreceptors (Kotrly et al. 1984), cerebral perfusion is better protected in the event of sudden hypovolaemia from blood loss, or induced hypovolaemia from diuretic agents.

Hypotension induced by isoflurane is due mainly to vasodilatation and to a lesser extent reduced cardiac output (Newberg, Milde, Michenfelder 1983). It is easily controlled by adjusting the delivered concentration of the drug. On decreasing the concentration of isoflurane the blood pressure slowly rises without rebound. Bishay 1984 found that tachyphylaxis did not occur even when hypotension was very prolonged.

Cerebral oxygen consumption is decreased by isoflurane up to one M.A.C. in cats (Todd, Drummond and Shapiro 1982), and dogs (Newberg, Milde, and Michenfelder 1983). The decrease in oxygen consumption that accompanies isoflurane anaesthesia may account for the ability of isoflurane to protect the brain against ischaemia under profound hypotension (Newberg and Michenfelder 1983).

Controlled Hypotension

Hypotension has long been used by neuroanaesthetists to produce a surgical field in which microsurgery is possible. This implies lowering the mean arterial blood pressure to 60 mmHg (8.0 kPa) for "moderate" hypotension and occasionally dropping the mean blood pressure to lower levels (profound hypotension) for short periods, for example, to enable the surgeon to secure the neck of a bleeding aneurysm. Hypotension is also of great assistance to the surgeon when excising vascular tumours and angiomas, the supplying vessels of which may have lost their ability to autoregulate, and do not constrict in response to hypocapnia. Hypotension thus reduces operating time, blood loss with risks of blood replacement, and decreases post-operative complications and morbidity. Hypotension is essential for most aneurysm surgery, supra- and infra-tentorial tumour surgery and transnasal operations.

Hypotension is attended by risks in some patients particularly if autoregulation is impaired by vascular spasm, raised intracranial pressure or hypertension, and in patients with generalized arteriopathy. Hypotension demands intraoperative monitoring using intra-arterial catheterisation also enabling arterial blood gas analysis, an intravenous line, and if necessary a central venous line. The end tidal CO_2 must also be monitored. A patient who has received muscle relaxants and is ventilated, rarely shows ischaemic changes in the electrocardiogram (E.C.G.).

During controlled hypotension it is our practice to use an increased inhaled oxygen concentration, a 50:50 oxygen, nitrous oxide mixture. Hypotension may lead to uneven pulmonary perfusion with consequent increase in pulmonary deadspace (Eckenhoff et al. 1963). This does not occur if cardiac output and right heart function is maintained. Some hypotensive agents, *e.g.* nitroprusside increase intrapulmonary shunting and therefore an oxygen concentration of 50% is helpful. The pharmacology and limitations of the various hypotensive agents must be clearly understood and selected to suit the practice of the surgeon and the medical state of the patient. Many drugs produce safe adequate hypotension, none of which need large dosage against a background of the mild hypotension of beta blockade.

Beta Blocking Drugs

Beta-adrenoreceptor blockade is a rational prelude to using directly acting vasodilator drugs (see below) or ganglion blockers, which can cause marked tachycardia. The drugs available are practolol and labetolol. Practolol is a competitive catecholamine antagonist at beta-adrenoreceptors, acting mainly on the beta receptor sites of cardiac muscle. Labetolol has both alpha and beta adrenoreceptor blocking activity, beta blockade predominating over alpha in the ratio of 5 : 1, the beta effect lasting up to 90 minutes. The dosage of labetolol varies in the range of 5 mg–200 mg according to body size, age and physical condition. The dose easily controlled as it acts synergistically with halothane, and is reversed by atropine. Hypotension is decreased by reducing the volatile agent.

Alpha Blocking Drugs

Alpha blockade may be needed when hypertension or normotension is resistant to beta blockade; alpha blockers such as phentolamine will act synergistically with the beta blockers already given. Alpha blockers are the drugs of choice when dealing with the hypertension caused by catecholamine secreting tumours encountered in neurosurgery such as glomus tumours invading the posterior fossa.

The alpha blocker phentolamine is a short acting beta-imidazoline, a powerful vasodilator having a mild sympathomimetic effect. It therefore lowers peripheral resistance and blood pressure, increasing cardiac output and heart rate. It decreases stroke volume and cardiopulmonary blood volume. Phentolamine can be given orally, allowing pre-operative control of blood pressure.

Trimetaphan

Trimetaphan is a short acting ganglion blocking agent lasting no more than 20 minutes. When the patient is positioned with head up tilt its principal action of lowering the peripheral resistance may be accentuated. A compensatory tachycardia is often seen, so that beta blockade with labetolol or practolol should precede the use of trimetaphan. The dose needed to produce a systolic blood pressure of 80 mmHg may be as little as 2.5 mg, this can always be increased by additional boluses. Trimetaphan releases histamine in susceptible subjects and a history of any previous episode of asthma precludes its use, just as it precludes the use of non-selective beta blockade. A comparison of trimetaphan with sodium nitroprusside (SNP) has shown that trimetaphan has minimal effect on I.C.P. but does not maintain the microcirculation as well as SNP (Ishikawa 1982, Ishikawa and McDowall 1980).

Direct Acting Vasodilators—Sodium Nitroprusside and Nitroglycerine

SNP is the most powerful hypotensive agent and has the virtue of predictable effectiveness in all patients. Its use however, is not surprisingly associated with difficulties which demand very close and continuous supervision. It is mainly used with patients in whom other agents are contraindicated (*e.g.* asthmatics) or ineffective as may happen with the youthful and physically fit.

Sodium nitroprusside acts as a calcium antagonist, by interfering with the mobilisation of intracellular Ca^{++}. S.N.P. contains 5 cyanide radicals, which are liberated on breakdown of the drug. 98% of the cyanide produced is contained in red blood cells; a small proportion is combined with either methaemoglobin or Vitamin B_{12}. Cyanide is metabolised by liver rhodanase to thiocyanate but sulphydryl groups must be available. Sodium thiosulphate can be given to supply them and reduce cyanide levels (Krapez et al. 1981) and does not affect the hypotension.

S.N.P. is given as a continuous infusion in 5% dextrose, 50 mg in 500 ml of solution, regulated to maintain the required mean blood pressure; a constant infusion pump, or other mechanical regulator must be used. The infusion should be given through a cannula sited for the purpose. Preference is for an arm vein as there is a lag period if a lower limb vein is used. The patient is usually in a position of fairly steep head up tilt, and vasodilation and pooling of venous blood occurs. Toxicity will be detected if the dose administered exceeds 3.5 mg/Kg (half the lethal dose). Blood gases must be measured as cyanide toxicity can be detected by falling pH and decreasing PaO_2 and alteration of the plasma lactate and lactate/pyruvate levels. When the base deficit falls to — 6 mmols/l the drug must be discontinued (Tinker and Michenfelder 1978).

Longer term infusion to control post-operative hypertension must not exceed 8 ug/Kg^{-1}/min^{-1} in the presence of adequate sulphhydryl groups (Tinker and Michenfelder 1976).

Brain oxygen tension is maintained in S.N.P. induced hypotension (Longnecker et al. 1980, Maekawa et al. 1979). Cerebral blood flow remains adequate even if cerebral perfusion pressure is reduced to low levels. Cerebral extracellular ionic haemostasis is better preserved than in trimetaphan induced hypotension (Morris et al. 1983), but hypotension induced by isoflurane even more closely preserves ionic homeostasis (Morris and Cooper 1987).

SNP is a vasodilator directly affecting cerebral vessels increasing intracranial blood volume (Tinker et al. 1977). It should therefore only be used after the dura is open, and during hyperventilation and the administration of 50% oxygen, particularly in patients with low intracranial compliance (Cottrell et al. 1980).

Withdrawal of SNP is followed by rapid rise of blood pressure, often to levels higher than the control level. This rebound hypertension is one of the great disadvantages of sodium nitroprusside as the small vessels of the operated area are unprotected by autoregulation; bleeding and/or oedema may occur. This phenomenon is less often seen in the patient who is well beta blocked, as it is due to the effect of SNP on the renin angiotension system, with increase of plasma catecholamines and plasma renin activity (Miller et al. 1977). In my opinion beta blockade in necessary in all patients in whom it is not positively contra-indicated, *e.g.* asthmatic subjects. It must be remembered that the pulmonary circulation can also be affected by rebound hypertension, resulting in pulmonary oedema and hypoxaemia (Shibutami et al. 1983).

Nitroglycerine produces a steadier and less precipitate fall in blood pressure. Recovery is slower than with nitroprusside and can be prolonged. It is unpredictable in its ability to produce hypotension when needed. In one series hypotension was not produced in three out of twenty two patients (Chestnut and Albin 1978). It is therefore not popular with neuroanaesthetists.

Hydralazine given intravenously will produce hypotension but there is a delay of 10–15 minutes after a bolus dose given intravenously. This impairs its usefulness in acute situations though it can be useful in post-operative management.

Adenosine triphosphate and adenosine are potent vasodilators and may prove to be useful drugs but are not yet generally available. The induction of hypotension is rapid as is the eventual recovery. Bradycardia rather than tachycardia is produced (Sollevi, Langerkranser et al. 1984).

Post-operative control of blood pressure is often needed after surgery for excision of large vascular tumours and arteriovenous malformations, or if the patient previously suffered from poorly controlled hypertension. It is logical to continue in the post-operative period the hypotensive technique successful during the anaesthetic. In the majority of cases this means continuing a labetalol infusion. Labetalol 100 mg in 100 ml normal saline given at 80 drops/minute from a paediatric burette will control the blood pressure in most cases, but fluid overload must be avoided. If necessary hydralazine 10–20 mg can be added intramuscularly or intravenously, maximum doses of up to 500 mg in 24 hours can be used. It is also possible to continue a nitroprusside infusion 8 ug/Kg^{-1}/min^{-1}, continuing also the beta blockade, and intra-arterial pressure monitoring.

The Sitting Position and Air Embolism

A survey of all United Kingdom centres (Campkin 1981) found that 52% of centres used the sitting position regularly for posterior fossa exploration and 31% for cervical laminectomies. In the United States of America the figures are only slightly less (Tausk and Miller 1983).

The advantages to the surgeon of access to midline tumours, low venous pressure (and therefore little bleeding) and comfortable operating posture with the wound directly ahead, outweigh the disadvantages. The anaesthetist aims to maintain the systolic blood pressure at or near normal levels, to use controlled hypotension only if requested, and only to a systolic pressure of 100 mm/Hg. Above all, the anaesthetist must ensure that the central venous pressure is high enough to prevent entry of air into any inadvertently opened veins or sinuses. Monitoring of central venous pressure, arterial pressure by direct reading, and a continuous display of end tidal CO_2 are mandatory for all procedures. The central venous line must be positioned, using the anticubital fossa and brachial vein, so that the "T" of the E.C.G. becomes bifid, resembling a "W". The catheter will have its tip at the junction of the superior vena cava and right atrium. The position must be rechecked when the patient is sitting and the neck and shoulders are flexed. Prevention of air embolism dominates all anaesthetic techniques for posterior fossa surgery in this position.

After the C.V.P. is recorded from the long line, positioned as described, the patient should be given a fluid load, crystalloid, or plasma expander until the C.V.P. is at least positive, ideally above the level of the wound.

Following induction with an intravenous agent, and intubation with suxamethonium 1 mg/Kg, muscle relaxation is achieved choosing a further relaxant which will ensure cardiac stability. Vecuronium, atracurium by infusion, or pancuronium are all suitable and will produce little change in blood pressure or pulse rate: tubacurarine is often used in young patients. When the patient is placed in the sitting position, the systolic blood pressure level and thus cerebral perfusion pressure must be maintained; anaesthetic agents which are more powerful vasodilators and cardiac depressants are avoided.

Trichloroethylene has retained a place here as a useful anaesthetic and analgesic agent for old and frail patients, its analgesic effect lasting into the post-operative period. The cardiovascular system remains very stable. Isoflurane is gaining in popularity and can be used in these cases, particularly when combined with a short infusion of alfentanil or a small dose of fentanyl during the most stimulating parts of the operation, *i.e.*, the incision, muscle and bone work.

The anti-gravity suit is applied to all patients in this unit who undergo surgery in the sitting position. The suit is the aviation type (Hewer and Logue 1962, Fruenchen 1959). The combination of leg and abdominal compression at an inflation pressure of 60 mm/Hg prevents arterial hypotension, reduces the amount of intravenous fluids and blood transfused and greatly assists in reducing the frequency and severity of air embolism. In the operating theatre positive end expiratory pressure (P.E.E.P.) is used to assist in raising the venous pressure. The combination of P.E.E.P. at + 8 to 10 cm of water with the anti-gravity suit raises the venous pressure by 10 cms of water, but in some patients air embolism will still occur. Attempts to selectively compress the jugular veins in the neck with a tourniquet when air embolism is anticipated have so far not been successful. An air embolus can occur at any time during a suboccipital exploration, and must be differentiated from the similar cardiovascular effects of brain stem ischaemia, a fall in pulse rate and then blood pressure, usually caused by transitory spasm of the posterior inferior cerebellar artery, or manipulation of an intramedullary or lower pontine tumour. When the fall in blood pressure and pulse is due to surgical manoeuvres the drop in end tidal CO_2 is not sustained, though falling blood pressure leads to a continued slow fall in $paCO_2$. Diagnosis of air embolism depends on falling CO_2 excretion from the lungs, and is detected by an infra-red analyser capable of quantifying the dilution of CO_2 by air.

The precordial doppler probe potentially so useful in the detection of air embolism, is often unreliable. It must be correctly positioned over the atrium, and is very sensitive to interference from surgical diathermy, thus losing its capacity to alert the anaesthetist at the moment it is needed. The oesopharyngeal doppler probe may prove to be a more reliable instrument.

The treatment of air embolism must be immediate, the surgeon on being informed applies a large saline swab to the wound. Immediate compression of the neck veins by the anaesthetist may reveal to the surgeon the source of the air as blood and air escapes from the open wound, the offending vein is then easily secured. Nitrous oxide must be withdrawn from the anaesthetic circuit to minimise the expansion of air bubbles in the blood, the solubility of nitrous oxide being thirty times that of nitrogen, a large volume of nitrous oxide will replace the nitrogen of the air. Aspiration of the right arterial catheter may result in removal of some air. Hypotension following air embolism is a late result in this rapid time sequence. It is treated symptomatically with plasma expanders or blood transfusion. E.C.G. changes seen in a display of lead 1 are late developments with changes in P and T waves, right bundle branch block and S.T. segment depression. In severe cases these are followed by bradycardia and cardiac arrest. In the rare case of a patient failing to respond to immediate treatment, the patient must be repositioned preferably in the left lateral position and further attempts to aspirate air made; full cardiac resuscitation is carried out as necessary. There is little point in discontinuing an operation in a resuscitated patient who has suffered little or no anoxia. The operation should be continued in the lateral or prone position as appropriate, and I.P.P.V. continued for some time post-operatively, at least until the patient shows signs of returning consciousness and effective spontaneous respiration.

Cardiovascular Changes in Posterior fossa Exploration

Monitoring of the arterial pulse by direct cannulation is mandatory during surgery for posterior fossa tumours. Continuous display of the E.C.G., pulse rate and systolic and diastolic blood pressure will provide warning to the anaesthetist that a surgical manoeuvre is disturbing either directly the vasomotor centre beneath the calamus scriptoris of the 4th ventricle, or that brain stem ischaemia is at that moment being produced by manipulation of brain stem branches of posterior inferior cerebellar artery or anterior inferior cerebellar artery. In the first case of disturbance of the vasomotor centre a rising blood pressure is seen without great disturbance of the pulse rate (a rise of pulse rate of 10/minute is an average finding). This persistent rise in blood pressure is harmless if the precipitating surgical manoeuvre ceases; the blood pressure settles to its original level. Transient brain stem ischaemia demonstrates itself by bradycardia and occasionally asystole. Bradycardia responds immediately to atropine 0.3 mg–0.6 mg. Asystole requires atropine and swift cardiac massage, and if necessary full resuscitation. Interpreting the monitor in the light of the anaesthetic drugs is of vital importance, and presents problems to junior anaesthetists who unless restrained tend to treat symptoms and neglect or obscure the underlying cause. To enable the monitor to be read intelligently, the anaesthetic is adjusted to provide a continuously steady systolic blood pressure and pulse rate, adjusting the inhalational agent as necessary, avoiding opiates and their tendency to produce some hypotension, and beta blockers and therefore avoiding iatrogenic bradycardia. The changes described are reversible, particularly rapidly in the presence of hypocapnia. When combined with hypercapnia they are more sinister.

The Effect of Anaesthetic Drugs on Recordings of Evoked Potentials

Evoked potentials whether somatosensory, visual or auditory are specific for neuronal pathways, *e.g.* retina to occipital cortex. Measurements are made of latency, that time which elapses from the stimulus to the appearance of the potential (measured in milliseconds), and amplitude, from the first negative peak to the deepest following positive valley (measure in μV). Evoked potentials are used pre-operatively to test the integrity of specific neuronal pathways, and are changed by hypoxia variations in body temperature (Symon et al. 1979), interference with blood supply and tissue distortion from excessive retraction (Grundy 1983). A safe period of deliberate ischaemia, for example, the use of temporary clips in aneurysm surgery (Symon et al. 1984, Momma et al. 1987), can be timed by recording the changes in the recorded evoked potential. The effect of anaesthetic drugs on evoked potentials must be understood and allowed for in interpreting evoked potentials, the effects of anaesthetic drugs being present on both the operated and normal side of the brain. Each anaesthetic agent affects brain electrical activity and conduction differently. Intravenous and inhalational agents both prolong the latency and decrease the amplitude of cortical evoked potentials. At 0.5 M.A.C. or in the case of halothane 0.5% the central conduction time (C.C.T.) is increased without significant hypotension. At 1.5 M.A.C. hypotension is seen, and further prolongation of C.C.T. (Wang et al. 1985).

Halothane has been found to cause an increase in the latency of the visual evoked potential (V.E.P.) (Uhl, Squires, et al. 1980). Nitrous oxide up to 50% as commonly used in neuroanaesthesia produces a dose related reduction in V.E.P. and somato-sensory evoked potential (S.E.P.) amplitude without changing the latency (Sebel, Flynn, Ingram 1984). Isoflurane produces dose dependent increase in latency of S.E.P., V.E.P. and brain stem auditory evoked potential (B.A.E.R.). Amplitude reduction is seen in S.E.P., V.E.P. but B.A.E.R. are little affected (Sebel et al. 1986).

Drugs used in anaesthetic practice, alter consciousness and very variably affect later components of the somato-sensory evoked potential. Comparison of affected and normal sides of the body is essential to interpretation.

Benzodiazepines and opiates (fentanyl) are exceptions and can be used without affecting brain electrical response monitoring (Domino 1967, Samra et al. 1984).

Further studies must be carried out on the effect of hypocarbia and hypotension on evoked potentials.

Summary

Over the years the basic principles underlying the practice of neuroanaesthesia have not changed, but introduction of new anaesthetic agents and associated techniques have improved the ability of the neuroanaesthetist to "fine tune" the patients physiological state. This has improved the capacity of the neuroanaesthetist to mitigate the inevitable fluctuations which occur and prevent their ill effects. Further improvement is still desirable and possible. It takes years for the correct plan of usage of new drugs to be formulated for the clinical situation, and their relationships established to new techniques of patient monitoring. Like neurosurgery itself neuroanaesthesia shows no signs of approaching a final definitive state in the forseeable future.

References

1. Adams RW, Gronert GA, Sundt TM Jr, Michenfelder JD (1972) Halothane, hypocapnia and cerebrospinal fluid pressure in neurosurgery. Anesthesiology 37: 510–517

1a. Albrecht RF, Miletich DJ, Madelar LR (1983) Normalization of cerebral blood flow during prolonged halothane anaesthesia. Anaesthesiology 58: 26–31

2. Artru AA, Nugent M, Michenfelder JD (1982) Enflurane causes a prolonged and reversible increase in the rate of C.S.F. production in dogs. Anesthesiology 57: 255–260

3. Artru AA (1984) Isoflurane does not increase the rate of C.S.F. production in the dog. Anesthesiology 60: 93–197
4. Artru AA (1983) Effects of halothane and fentanyl on the rate of C.S.F. production in the dog. Anesthesia and Analgesia 62: 581
5. Albrecht RF, Miletich DJ, Madelar LR (1983) Normalisation of cerebral blood flow during prolonged halothane anaesthesia. Anesthesiology 58: 26–31
6. Bishay EG, Fukimaga AF, Echiverri E, Smagalski G, Ito H, Webb J (1984) Cardiovascular effects of prolonged isoflurane induced hypotension. Anesthesia and Analgesia 63: 187
7. Burchiel KJ, Stockard JJ, Calverley RR, Smith NT (1977) Relationship of pre- and post anesthetic E.E.G. abnormalities to enflurane-induced seizure activity. Anesthesia and Analgesia 56: 509–514
8. Campkin TV (1981) Posture and ventilation during posterior fossa and cervical operations. Current practice in the United Kingdom. Br J Anaesth 53: 881–884
9. Carter ML (1983) Bradycardia after the use of atracurium. Br Med J 287: 247–248
10. Chestnut JS, Albin MS, Gonzalez-Abola E, Newfield P, Maroon JC (1978) Clinical evaluation of intravenous nitroglycerine for neurosurgery. J Neurosurg 48: 704–711
11. Christensen MS, Hoedt-Rasmussen K, Lassen NA (1967) Cerebral vasodilatation by halothane anaesthesia in man and its potentiation by hypotension and hypercapnia. Br J Anaesth 39: 927–934
12. Cottrell JE, Gupta B, Turndorf H (1980) Induced hypotension. In: Cottrell JE, Turndorf H (eds) Anesthesia and neurosurgery. Mosby, St. Louis, pp 387–400
13. Cousins MJ, Mazze RI (1973) Methoxyflurane nephrotoxicity—a study of dose response in man. J Am Med Assoc 225: 1611–1616
14. Cohen PJ, Alexander SC, Smith TC (1967) Effect of hypoxia and normocarbia on cerebral blood flow and metabolism in conscious man. J Appl Physiol 23: 183
15. Domino EF (1967) Effect of pre anesthetic and anesthetic drugs on visually evoked responses. Anesthesiology 28: 184–191
16. Drummond JC, Todd MM, Shapiro HM (1983) Cerebral blood flow autoregulation in the cat during anesthesia with halothane and isoflurane. Anesthesiology 59: A 305
17. Eger EI 2nd, Smith NT, Stoelting RK, Cullen DJ, Kadish LB, Whintcher CE (1970) Cardiovascular effects of halothane in man. Anesthesiology 2: 396–409
18. Eckenhoff JE, Enderby GEH, Larson A, Edridge A, Judeuine DE (1963) Pulmonary gas exchange during deliberate hypotension. Br J Anaesth 35: 750–759
19. Fruenchen IP (1959) The use of an anti-gravity suit in neurosurgery. Acta Anaesthesiol Scand 3: 17–23
20. Grundy BL (1983) Intraoperative monitoring of sensory-evoked potentials. Anesthesiology 58: 72–87
21. Hewer AJH, Logue V (1962) Methods of increasing the safety of neuroanaesthesia in the sitting position. Anaesthesia 17: 476–481
22. Holaday DA, Fiserova-Bergerova V, Latto IP, Zubrbiel MA (1975) Resistance of isoflurane to biotransformation in man. Anesthesiology 43: 325–332

23. Ishikawa T, Funatsu N, Okamoto K, Takeshita H, McDowall DG (1982) Cerebral and systemic effects of hypotension induced by trimetaphan or nitroprusside in dogs. Acta Anesthesiol Scand 26: 643–648
24. Ishikawa T, McDowall DG (1980) Electrical activity of the cerebral cortex during induced hypotension with sodium nitroprusside and trimetephan in the cat. Br J Anaesth 53: 605–611
25. Jennett WB, Barker J, Fitch W, McDowall DG (1969) Effect of anaesthesia on intracranial pressure in patients with space-occupying lesions. Lançet 1: 61–64
26. Kotrly KJ, Ebert TJ, Vucins E, Igler FO, Barney JA, Kampine JP (1984) Baroreceptor reflex control of heart rate during isoflurane anesthesia in humans. Anesthesiology 60: 173–179
27. Krapez JR, Vesey CJ, Adams L, Cole PV (1981) Effects of cyanide antidotes used with sodium nitroprusside infusion: sodium thiosulphate and hydroxycobalamine given prophylactically to dogs. Br J Anaesth 53: 793–804
28. Lassen NA, Christensen MS (1976) Physiology of cerebral blood flow. Br J Anaesth 48: 719–734
29. Longnecker DE, Radke J, Schenk HD, Sonntag H (1980) Cerebral oxygenation during deliberate hypotension. Anesthesiology [Suppl] S 3: S 91
30. McDowall DG (1967) The effects of clinical concentrations of halothane on the blood flow and oxygen uptake of the cerebral cortex. Br J Anaesth 39: 186–196
31. Maekawa T, McDowall DG, Okuda Y (1979) Brain-surface oxygen tension and cerebral cortical blood flow during hemorrhagic and drug-induced hypotension in the cat. Anesthesiology 51: 313–320
32. Meyers RR, Shapiro HM (1978) Paradoxical effect of enflurane on cloroid plexus metabolism: clinical implications. American Society Anesthesiology 489
33. Miletich DJ, Ivankovich AD, Albrecht RF, Reinana R, Rosenberg R, McKissic EP (1976) Absence of autoregulation of cerebral blood flow during halothane and enflurane anesthesia. Anesthesia and Analgesia 55: 100–109
34. Miller ED Jr, Ackerley JA, Vaughan ED Jr, Peach MJ, Epstein RM (1977) The renin-angiotensin system during controlled hypotension with sodium nitroprusside. Anesthesiology 47: 257–262
35. Momma F, Wang AD, Symon L (1987) Effects of temporary arterial occlusion on somatosensory evoked responses in aneurysm surgery. Surg Neurol 27: 343–352
36. Morris PJ, Heuser D, McDowall DG, Hashiba M, Myers D (1983) Cerebral cortical extracellular fluid H^+ ions and K^+ activities during hypotension in cats. Anesthesiology 59: 10–18
37. Morris PJ, Cooper A (1987) Regional cerebral blood flow and cerebral cortical ion haemostasis during induced hypotension with isoflurane and sodium nitroprusside. Br J Anaesth (in press)
38. Moss E, Dearden NM, McDowall DG (1983) Effects of 2% enflurane on intracranial pressure and cerebral perfusion pressure. Br J Anaesth 55: 1083–1088
39. Murphy F, Kennel EM, Johnstone RE (1974) The effects of enflurane, isoflurane and halothane on cerebral blood flow and metabolism in man. Abstracts of Scientific Papers, Annual meeting of the American Society of Anesthesiologists 61

40. Neigh JL, Garman JK, Harp JR (1971) The electroencephalographic pattern during anesthesia with ethrane, effects of depth of anesthesia, $PaCO_2$ and N_2O. Anesthesiology 35: 482–487
41. Newberg LA, Milde JH, Michenfelder JD (1983) Canine systemic and cerebral effects of hypotension induced by isoflurane. Anesthesiology 59: A 17
42. Newberg LA, Milde JH, Michenfelder JD (1984) Systemic and cerebral effects of isoflurane-induced hypotension in dogs. Anesthesiology 60: 541–546
43. Newberg LA, Michenfelder JD (1983) Cerebral protection by isoflurane during hypoxemia or ischemia. Anesthesiology 59: 29–35
44. Neuberger J, Gimson AE, Davis M, Williams R (1983) Specific serological markers in the diagnosis of fulminant hepatic failure associated with halothane anaesthesia. Br J Anaesth 55: 15–19
45. Payne JP, Hughes R (1981) Evaluation of atracurium in anesthetised man. Br J Anaesth 53: 45–54
46. Sadove MS, Wallace VE (1962) Halothane. Blackwell, Oxford
47. Samra SK, Lilly DJ, Rush NL, Kirs MM (1984) Fentanyl anesthesia and human brain-stem auditory evoked potentials. Anaesthesiology 61: 261–265
48. Sebel PS, Flynn PJ, Ingram DA (1984) Effect of nitrous oxide on visual, auditory and somatosensory potentials. Br J Anaesth 56: 1403–1407
49. Sebel PS, Ingram DA, Flynn JP, Rutherfoord CF, Rogers H (1986) Evoked potentials during isoflurane anaesthesia. Br J Anaesth 58: 580–585
50. Shibutami K, Komatsu T, Bhalodia R, Kubal C, Taddonio R (1983) Rebound hemodynamic responses after the withdrawal of nitroprusside in anesthetized patients. Anesthesiology 59: A 16
51. Skinhoj E, Strandgaard S (1973) Pathogenesis of hypertensive encephalopathy. Lancet 1: 461–462
52. Sollevi A, Lagerkranser M, Irestedt L *et al* (1984) Controlled hypotension with adenosine in cerebral aneurysm surgery. Anesthesiology 61: 400–405
53. Strandgaard S, Olsen J, Skinhoj E (1973) Autoregulation of brain circulation in severe arterial hypertension. Br Med J 1: 507–510
54. Symon L, Hargadine J, Cone JD, Zawirski M, Wang AD, Watson A (1981) Monitoring of evoked electrical activity in cerebral ischaemia. In: Meyer JS, Lechner H, Reivich M, Ott EO, Aranibar A (eds) Cerebral vascular disease 3. Excerpta Medica, Amsterdam Oxford Princeton, pp 309–314
55. Symon L, Hargadine J, Zawirski M, Branston MN (1979) Central conduction time as an index of ischaemia in subarachnoid haemorrhage. J Neurol Sci 44: 95–103
56. Symon L, Wang AD, Costa E, Silva IE, Gentili F (1984) Per-operative use of somatosensory evoked responses in aneurysm surgery. J Neurosurg 60: 269–275
57. Tausk HC, Miller R (1983) Anesthesia for posterior fossa surgery in the sitting position. Bull NY Acad Med 59: 771–783
58. Tinker JH, Michenfelder JD (1978) Cardiac cyanide toxicity induced by nitroprusside in the dog: potential for reversal. Anesthesiology 49: 109–116
59. Tinker JH, Michenfelder JD (1976) and (1977) Sodium nitroprusside pharmacology, toxicology and therapeutics. Anesthesiology 45: 340–354
60. Todd MM, Drummond JC, Shapiro HM (1982) Comparative cerebrovascular

and metabolic effects of halothane, enflurane, and isoflurane. Anesthesiology 57: A 332
61. Uhl RR, Squires KC, Bruce DL, Starr A (1980) Effect of halothane anesthesia on the human cortical visual evoked response. Anesthesiology 53: 273–276
62. Vickers MD, Schnieden H, Wood-Smith FG (1984) Drugs in anaesthetic practice, 6th ed, General anaesthetics, Chapter 4, pp 140. Butterworths, London
63. Wang AD, Costa E Silva I, Symon L, Jewkes D (1985) The effects of halothane on somatosensory and flash visual evoked potentials during operations. Neurol Res 7: 58–62
64. Watkins J (1986) Histamine release and atracurium. Br J Anaesth 58 [Suppl 1]: 19S–22S
65. Williams R (1986) Halothane and the liver:—A medical review. Proceedings of a symposium on halothane and the liver. Univ Bristol Medical School, pp 12–19

Editorial Note

This chapter reflects what could be regarded as one of the principal techniques of neurosurgical anaesthesia in the United Kingdom at the present time. However, not all anaesthetists are entirely in favour of volatile halogenated anaesthesia and many units in the United Kingdom and abroad perform neurosurgery under gas and oxygen anaesthesia supplemented by techniques of neuroleptanalgesia with such drugs as phenoperidine and haloperidol. Many anaesthetists in France have abandoned trichlorethylene for many years and many anaesthetists in various parts of Europe are still attached to thiopentone anaesthesia.

This chapter, therefore, must be regarded as an acceptable technique, but not the only one.

Controversial Views of the Editorial Board Regarding the Management of Non-Traumatic Intracerebral Haematomas

(excluding those in the brain stem or arising from vascular malformation)

Collected by Professor Bernard Pertuiset. Edited by the Chief Editor

This chapter contains a review of the present status of the management of non-traumatic intracerebral haematoma in Europe. The questionnaire was addressed to the Editorial Board of Advances and Technical Standards and the views expressed represent a consensus of major interests throughout Europe.

1. Was the average time of admission to a neurosurgical service after haemorrhagic stroke?

In almost all the services covered, the average time of admission was several days after the ictus, although in the case of Cohadon severe cases would be admitted on the day of bleeding and the same applied to a number of cases from Guidetti and Nornes. In the majority of incidences, however, cases with intracerebral haematoma would be admitted first ot a medical service and subsequently transferred.

2. Are patients carried to hospital in an ambulance with special medical facilities?

These were available in Belguim according to Brihaye, in France according to Pertuiset and Cohadon, in Norway according to Nornes, mostly in Germany according to Loew and also in Zurich according to Yaşargil. In the services of Miller, Symon, Pásztor, specially medically equipped ambulances were not generally employed.

3. With the patient in coma is early tracheal intubation advised by the ambulance staff with artifical ventilation if necessary? Is medical treatment carried out on the ambulance?

There was almost general agreement that comatose patients should be transferred at least intubated and in Brihaye's case all such admissions were transferred at once to a intensive care unit. Most of the editors believed that no specialized treatment should precede CT examination or a definite diagnosis but is interesting that Professor Loew would immediately start anticonvulsive prophylaxis and nimodipine which would be stopped if subsequent investigations showed no blood in the subarachnoid space. Yaşargil suggested that dexamethasone would be given by the majority of primary care physicians before the patient was transferred. Both Pásztor and Symon pointed out that transfer or partially conscious patients should be carried out in the lateral semi-prone position and neither advocated intubation by ambulance staff.

4. What routine investigations are employed in the case of supra-tentorial haematoma?

It has become obvious from the replies of all the editors that CT scan is the key examination. Brihaye stressed the importance of the routine examinations of an intensive care unit, ECG, EEG, chest and skull X-rays, blood gas determination, and haematological workup but all agreed that only in the case of suspicion of aneurysm would angiography be carried out. In all the major centres of Europe, CT and angiography are evidently available now, day or night. All the authors stressed the importance of clinical examination by a competent surgeon.

5. Should a lumbar puncture be performed?

There was uniform agreement that lumbar punctures were not necessary and should not be performed, it has thus become clear that the primary investigation in the verification of an intracranial haemorrhage should be CT, and several of the editors stressed the unfortunate results of potential lumbar punctures performed outside. This perhaps contrasts with the opinions expressed in the questionnaire on the management of subarachnoid haemorrhage in which where the patient was fully conscious and had neck stiffness suggestive of subarachnoid haemorrhage, lumbar punctures were still often performed in primary care centres in Europe before referral to a neurosurgical unit.

6. In which case is early surgery, i.e., on 1st or 2nd day advised? Does this advice depend on the clinical condition or investigations or both?

The only consensus for early operation would be in the case of lobar haematomas. Loew gave the criteria that lobar haematomas with a midline shift of more than 0.5 cms should be operated upon and the same point was made by Pertuiset stressing particularly that temporal haematomas in the young patient should be evacuated lest temporal herniation supervene. Cohadon proposed early surgery in patients who were in coma or whose clinical situation was rapidly deteriorating but also commented that the location of the haematoma influenced him and a similar reply was received from Guidetti who excepted patients with extensive basal ganglia haemorrhage, Pertuiset used the Hunt and Hess grading to determine that patients over the age of 65 and Hunt and Hess grade 4, 5, would not be operated upon and Loew also excluded operation in poor general condition or a biologically age of more than 70 years. Symon stressed that where evacuation of a polar frontal or temporal tip haematoma or an external capsular haematoma was to be considered angiography should precede surgery.

7. If early surgery is to be performed is this done through a craniotomy, craniectomy or a burr hole aspiration?

Interestingly here, only Cohadon used a burr hole initially. He would aspirate the haematoma till the ICP returned to a normal value and remove the haematoma totally only when ICP rose again to over 25–30 mm of mercury. All the other editors preferred a more generous approach either through a large trephine or small bone flap although Loew commented that where a clot had not been removed early its subsequent lysis might permit burr hole evacuation. He also commented that in his service large bone flaps were removed and used as a model for acrylic cranio-plastic later, and that they also enlarged the dural sac with a pericranial flap. Several authors, Guidetti and Cohadon, clearly relied on intracranial pressure measurments as a guide. Pásztor point out that the precision of pre-operative localization on CT had enabled evacuation under direct vision through relatively small exposures.

8. If early surgery is performed are the clots partially or totally removed?

Cohadon pointed out that the majority of deep haematomas in the thalamus or putamen were not operated on anyway so the clot would not be completely removed, but most other authors commented that if one had gone to the bother of exposing the clot then it should be removed as completely as possible though Miller cautioned that in early surgery one should be careful not to start further bleeding. Both Pertuiset and Pásztor commented that incomplete removal of the haematoma was preferable to starting excessive bleeding, and Symon commented that the use of self-retaining retraction, mirrors and the headlight enabled very thorough removal of the clots which was also Yaşargil's policy.

9. Do indications and technique vary with the location of the haematoma?

In general, most of the editorial board would not operate on deep haematomas in the dominant hemisphere and both Nornes and Pásztor commented that the expected results of the evacuation would modify both the decision to operate and the extent of the evacuation.

10. If there is blood in the ventricular system does this alter the indication for surgery and is external ventricular drainage performed?

Cohadon commented that intraventricular blood was no contra-indication to surgery as far as he was concerned and he used ventricular drainage from time to time. Both Brihaye and Guidetti used intra-ventricular drainage occasionally and Loew commented that large haematomas frequently ruptured into the ventricles and this did not influence his policy for or against operation, but that external ventricular drainage might be necessary if there was a great deal of blood in the ventricle. Symon commented that in his service, external ventricular drainage was never employed acutely, but could be used later if communicating hydrocephalus appeared, in which case a ventriculo-peritoneal shunt would usually be performed. It is interesting that the consensus appeared in favour of ventricular drainage if there was a great deal of blood in the ventricle and that this did not really influence the choice for or against surgery which was made on other grounds.

11. Does the indication for early surgery vary with associated disease with age or with anticoagulant therapy?

Loew stressed the importance of biological age and the majority of editors regarded extreme age and severe associated disease not necessarily as influencing the immediate management but certainly modifying the extent of subsequent treatment. All indicated that coagulopathy must be reversed before operation and Miller noted that arterial hypotension was best managed after the evacuation of the haematoma and cautioned against the lowering of arterial pressure in the presence of mass lesion and raised intracranial pressure.

12. When is early surgery contraindicated?

None of the editors would operate on a verified intracerebral haemtaoma in an aged patient, in a moribund condition, Miller summarized the views in the words "with signs of brain stem dysfunction I do not recommend surgery nor do I recommend surgery for patients with deep dominant hemisphere haematomas associated with massive neurological deficit including complete aphasia". All editors stressed the importance of reversing anticoagulants before surgery.

13. When early surgery is not performed what kind of monitoring and treatment do you advise?

Cohadon, monitored the intracranial pressure for two to six days, Nornes used intracranial pressure monitoring and monitoring with the cerebral function monitor with if necessary "barbiturate coma" Loew also used barbiturate sedation with hyperventilation and intracranial pressure registration but the others considered the management of an intensive care unit sufficient without intracranial pressure monitoring. Several, *e.g.*, Pásztor and Guidetti would use mannitol though this was clearly not a general view, and there was similar disagreement or failure to mention the use of dexamethasone. Those who did use it were unconvinced by its value.

14. When the patient has not been operated on in the early stage what would be your indications, if any, for delayed surgery?

Guidetti commented that delayed surgery was indicated because of late deterioration and this appeared to be a fairly general view. Evidence of enlargement of the haematoma on CT (Pertuiset) or delayed transfer of the patient from another unit with a lobar haematoma (Symon) seem to be further reasons for delayed operation and in Yaşargil's view a late operation was needed and would be indicated only if angiographic studies have revealed an arterio-venous malformation.

15. In a patient, who has survived, and is conscious with a severe neurological deficit should liquid haematoma be removed and when?

Similar answers were received here to question 14, midline shift according to Brihaye, advancing intracranial pressure according to Guidetti, persistent drowsiness or headache according to Miller. None were terribly convinced that the effectiveness of this manoeuvre and Symon pointed out that there was really no evidence that removal of a liquid haematoma late, accelerated convalescence.

16. Is angiography performed before the patient is discharged?

The general consensus was that angiography would be performed only in the young patients who were hypotensive and in patients with subcortical or lobar haematomas. Nornes and Symon pointed out that angiography would be performed any operative procedure and several of the editors indicated that were there was a haematoma in an unusual situation angiography must be performed to exclude an unusual aneurysm or arterio-venous malformation.

Cerebellar Haematomas

17. What is their frequency with respect to the supratentorial haematomas?

Figures between 1 and 8% emerged from the editorial board but it was clear that since regular CT scanning had been employed, posterior fossa haematomas were being increasingly discovered and in Pásztor's view the ratio of supra tentorial to cerebellar haematomas at the present time stood at 6 : 1.

18. How do you establish the diagnosis?

This was obtained invariably done by CT scanning with an appropriate clinical picture and Symon, Miller and Pertuiset commented on the utility of angiography especially when operation was being considered.

19. Do you perform early surgery or only external ventricular drainage?

Here Guidetti and Symon both pointed out that small haematomas diagnosed on CT could well be managed conservatively, provided the patient was conscious and was not suffering from raised ICP. Brihaye and others pointed out that external ventricular drainage alone was unsatisfactory the haematoma should be removed and drainage added if necessary and although Loew would drain the ventricles in larger haematomas with some degree of disturbance of consciousness and ventricular dilatation, he would operate on the haematoma if deterioration followed drainage. Miller pointed out that in his experience ventricular drainage made a small number of patients worse.

20. In the case of early surgery which position should be used for operation?

The majority of authors favoured supine or "park bench" position, most agreeing with Brihaye and Symon that while younger patients would sustain the sitting position, older patients should be operated upon lying down. Only Pásztor felt that the sitting position presented such considerable advantages in terms of surgical access and time that he would always employ if it was at all possible. Yaşargil apparently used the sitting position but made no comment on its specific indications.

21. Should the whole haematoma be removed or only a part of it?

There was uniform agreement that the whole haematoma should be removed.

22. Are these haematomas more severe than the supratentorial ones?

The old view of the dangers of posterior fossa haematomas were stressed by Cohadon and Guidetti, pointing out that patients in poor clinical condition or diagnosed late with large posterior fossa haematomas very rapidly died without surgery. It was clear, however, that modern CT scan diagnosis had produced evidence of small intra-tentorial haematomas which had an excellent prognosis and could be managed conservatively with care to observe the ventricular size, the general agreement that large infratentorial haematomas were undoubtedly more dangerous than the supratentorial equivalent, and should be removed.

Author Index

Subject Index

Druck: Adolf Holzhausens Nfg., Universitätsbuchdrucker

Advances and Technical Standards in Neurosurgery

Volume 1

1974. 96 figures. XI, 210 pages.
Cloth DM 102,—, öS 714,—
ISBN 3-211-81218-0

Contents:

Volume 2

1975. 150 partly coloured figures. XI, 217 pages.
Cloth DM 113,—, öS 790,—
ISBN 3-211-81293-8

Contents:

Volume 3

1976. 77 figures. XI, 154 pages.
Cloth DM 92,—, öS 644,—
ISBN-3-211-81381-0

Contents:

Advances and Technical Standards in Neurosurgery

Volume 4

1977. 66 partly coloured figures. XI, 154 pages.
Cloth DM 92,—, öS 644,—
ISBN 3-211-81423-X

Contents:

Advances: N. A. Lassen and D. H. Ingvar: Clinical Relevance of Cerebral Blood Flow Measurements. — G. W. Taylor and J. S. P. Lumley: Extra-Cranial Surgery for Cerebrovascular Disease. — J. Rétif: Intrathecal Injection of a Neurolytic Solution for the Relief of Intractable Pain.

Technical Standards: L. Symon: Olfactory Groove and Suprasellar Meningiomas. — M. G. Yaşargil, R. D. Smith, and J. C. Gasser: Microsurgical Approach to Acoustic Neurinomas. — G. Debrun, P. Lacour, and J. P. Caron: Balloon Arterial Catheter Techniques in the Treatment of Arterial Intracranial Diseases.

Volume 5

1978. 78 figures. XII, 224 pages.
Cloth DM 123,—, öS 860,—
ISBN 3-211-81441-8

Contents:

Advances: A. M. Landolt: Progress in Pituitary Adenoma Biology. Results of Research and Clinical Applications. — J. Hildebrand and J. Brihaye: Chemotherapy of Brain Tumours. — S. Mingrino: Supratentorial Arteriovenous Malformations of the Brain.

Technical Standards: J. Hankinson: The Surgical Treatment of Syringomyelia. — F. Loew and W. Caspar: Surgical Approach to Lumbar Disc Herniations. — B. Pertuiset, D. Fohanno, and O. Lyon-Caen: Recurrent Instability of the Cervical Spine With Neurological Implications—Treatment by Anterior Spinal Fusion.

Volume 6

1979. 79 figures. XI, 191 pages.
Cloth DM 113, —, öS 790,—
ISBN 3-211-81518-X

Contents:

Advances: E.-O. Backlund: Stereotactic Radiosurgery in Intracranial Tumors and Vascular Malformations. — J. Klastersky, L. Kahan-Coppens, and J. Brihaye: Infection in Neurosurgery. — C. Gros: Spasticity-Clinical Classification and Surgical Treatment.

Technical Standards: P. J. Derome and G. Guiot in co-operation with B. Georges, M. Porta, A. Visot, and S. Balagura: Surgical Approaches to the Sphenoidal and Clival Areas. — R. Braakman: Cervical Spondylotic Myelopathy. — F. Isamat: Tumours of the Posterior Part of the Third Ventricle: Neurosurgical Criteria.

Volume 7

1980. 147 figures. XI, 247 pages.
Cloth DM 125,—, öS 875,—
ISBN 3-211-81592-9

Contents:

Advances: M. G. Yaşargil, R. W. Mortara, and M. Curcic: Meningiomas of Basal Posterior Cranial Fossa.

Technical Standards: A. M. Landolt and P. Strebel: Technique of Transsphenoidal Operation for Pituitary Adenomas. — Surgical Treatment of Facial Nerve Paralysis; Longterm Results: H. Millesi: Extratemporal Surgery of the Facial Nerve — Palliative Surgery. S. Mingrino: Intracranial Surgical Repair of the Facial Nerve. U. Fisch: Management of Intratemporal Facial Palsy.

Volume 8

1981. 135 partly coloured figures. XII, 328 pages.
Cloth DM 157,—, öS 1100,—
ISBN 3-211-81665-8

Contents:

Advances: E. de Divitiis, R. Spaziante, and L. Stella: Empty Sella and Benign Intrasellar Cysts. — B. Pertuiset, D. Ancri, and A. Lienhart: Profound Arterial Hypotension (MAP ≦ 50 mm Hg) Induced with Neuroleptanalgesia and Sodium Nitroprusside (Series of 531 Cases). Reference to Vascular Autoregulation Mechanism and Surgery of Vascular Malformations of the Brain. — F. Gullotta: Morphological and Biological Basis for the Classification of Brain Tumors. With a Comment on the WHO-Classification 1979.

Technical Standards: E. Pásztor: Surgical Treatment of Spondylotic Vertebral Artery Compression. — P. Harris, I. T. Jackson, and J. C. McGregor: Reconstructive Surgery of the Head. — A. N. Konovalov: Operative Management of Craniopharyngiomas.

Volume 9

1982. 88 figures. XI, 177 pages.
Cloth DM 102,—, öS 714,—
ISBN 3-211-81718-2

Contents:

Advances: K. Faulhauer: The Overdrained Hydrocephalus. Clinical Manifestations and Management. — A. P. Romodanov and V. I. Shcheglov: Intravascular Occlusion of Saccular Aneurysms of the Cerebral Arteries by Means of a Detachable Balloon Catheter. — H. Spiess: Advances in Computerized Tomography.

Technical Standards: L. Symon: Surgical Approaches to the Tentorial Hiatus. — F. Loew: Management of Chronic Subdural Haematomas and Hygromas. — B. Williams: Subdural Empyema.

Volume 10

1983. 70 figures (1 in color). XI, 231 pages.
Cloth DM 123,—, öS 860,—
ISBN 3-211-81750-6

Contents:

Advances: R. J. S. Wise, G. L. Lenzi, and R. S. J. Frackowiak: Applications of Positron Emission Tomography to Neurosurgery. — J. Siegfried and T. Hood: Current Status of Functional Neurosurgery. — B. Pertuiset, D. Ancri, J. P. Sichez, M. Chauvin, E. Guilly, J. Metzger, D. Gardeur, and J. Y. Basset: Radical Surgery in Cerebral AVM — Tactical Procedures Based upon Hemodynamic Factors.

Technical Standards: M. Sindou and A. Goutelle. Surgical Posterior Rhizotomies for the Treatment of Pain. — A. Kumar and U. Fisch: The Infratemporal Fossa Approach for Lesions of the Skull Base.

Advances and Technical Standards in Neurosurgery

Volume 11

1984. 1 portrait. 80 figures. XII, 248 pages.
Cloth DM 125,—, öS 875,—, ISBN 3-211-81806-5

Contents:

Volume 12

1985. 49 partly colored figures. XI, 186 pages.
Cloth DM 114,—, öS 798,—, ISBN 3-211-81877-4

Contents:

Volume 13

1986. 77 partly colored figures. IX, 179 pages.
Cloth DM 114,—, öS 798,—, ISBN 3-211-81885-5

Contents:

Volume 14

1986. 71 partly colored figures. XIII, 230 pages.
Cloth DM 148,—, öS 1040,—, ISBN 3-211-81930-4

Contents:

Springer-Verlag Wien New York

Mölkerbastei 5, A-1011 Wien · 175 Fifth Avenue, New York, NY 10010, USA
Heidelberger Platz 3, D-1000 Berlin 33 · 37-3, Hongo 3-chome, Bunkyo-ku, Tokyo 113, Japan

MIX
Papier aus verantwortungsvollen Quellen
Paper from responsible sources
FSC® C105338

If you have any concerns about our products,
you can contact us on
ProductSafety@springernature.com

In case Publisher is established outside the EU,
the EU authorized representative is:
Springer Nature Customer Service Center GmbH
Europaplatz 3, 69115 Heidelberg, Germany

Printed by Libri Plureos GmbH
in Hamburg, Germany